The Role of Selenium in Health and Disease

The Role of Selenium in Health and Disease

Special Issue Editors

Catherine Méplan
David J. Hughes

MDPI • Basel • Beijing • Wuhan • Barcelona • Belgrade • Manchester • Tokyo • Cluj • Tianjin

Special Issue Editors

Catherine Méplan
Newcastle University
UK

David J. Hughes
University College Dublin
Ireland

Editorial Office
MDPI
St. Alban-Anlage 66
4052 Basel, Switzerland

This is a reprint of articles from the Special Issue published online in the open access journal *Nutrients* (ISSN 2072-6643) (available at: https://www.mdpi.com/journal/nutrients/special_issues/ Role_Selenium_Health_Disease).

For citation purposes, cite each article independently as indicated on the article page online and as indicated below:

LastName, A.A.; LastName, B.B.; LastName, C.C. Article Title. *Journal Name* **Year**, *Article Number, Page Range.*

ISBN 978-3-03936-146-5 (Hbk)
ISBN 978-3-03936-147-2 (PDF)

Contents

About the Special Issue Editors

Catherine Méplan (Lecturer in genetics). Dr Méplan has worked at Newcastle University, UK since 2003. Her research interests focus on the application of nutrigenomics technologies to the understanding of the relationship between selenium status and various cancers. She has contributed to the discovery of the importance of genetic variations in selenoprotein genes on selenium bioavailability, as well as using combined transcriptomics and proteomics approaches to understand how selenium status affects molecular pathways involved in cancer development. She and Dr Hughes are currently collaborating on several European projects looking at the effects of interactions between inherited genetic variations and selenium status on disease risk. She is a member of the European Nutrigenomics network NUGO, and has worked in collaboration with many research groups on selenium research. Previously, she has worked on the p53 tumour suppressor protein and cancer at the Biomedical Research Centre (Dundee University, UK), and during her PhD in Biology at the WHO/International Agency for Research on Cancer (IARC, Lyon, France). She received her BSc and MSc in Molecular Biology from the University of Grenoble, France.

David J. Hughes (Assistant Professor of Cancer Epidemiology). Dr Hughes is a Fellow of the Conway Institute in University College Dublin (UCD), Ireland. He currently leads several international multidisciplinary cohort studies of nutritional, genetic & microbial cancer molecular epidemiology. Dr Hughes is a PI in the European Prospective Investigation of Cancer and Nutrition (EPIC) and a member of its Gastrointestinal Cancer Working Group, a governing council member of the International Society for Selenium Research, and a member of the International Cancer Microbiome Consortium. Previously he has worked at the WHO/International Agency for Research on Cancer (IARC, Lyon, France), the Sanger Institute (Cambridge, England), and Imperial College (London, England). Dr Hughes received his BSc Biochemistry degree from the University of Leeds (England), his PhD in Medical Genetics from Queen's University Belfast (Northern Ireland), and a Postgraduate Diploma in Health Professions Education from the RCSI (Dublin, Ireland).

Editorial

The Role of Selenium in Health and Disease: Emerging and Recurring Trends

Catherine Méplan [1,*] and David J. Hughes [2]

[1] School of Biomedical, Nutritional and Sport Sciences, Newcastle University,
    Newcastle upon Tyne NE1 7RU, UK
[2] Cancer Biology and Therapeutics Group, UCD Conway Institute, School of Biomolecular and Biomedical
    Science, University College Dublin, D04 V1W8 Dublin, Ireland; david.hughes@ucd.ie
*   Correspondence: Catherine.Meplan@newcastle.ac.uk; Tel.: +44-(0)191-208-7664

Received: 1 April 2020; Accepted: 6 April 2020; Published: 10 April 2020

In this Special Issue of *Nutrients*, "The Role of Selenium in Health and Disease" covers diverse diseases in the 8 original research articles and 2 reviews, such as cardiovascular disorders (CVD), metabolic syndrome, obesity, cancer, and viral infection, and highlights novel potential biomarkers of disease risk and prognosis. From a public health perspective, the different manuscripts emphasize the already known U-shaped dose-response relationship between selenium (Se) concentration and disease risk across different study populations from Europe [1–3], the Middle East and North Africa [4], and Taiwan [5]. These reports therefore strengthen the importance of developing more personalized nutritional advice targeted at individuals at risk of disease with low Se intake. The results presented in this *Issue* also emphasize a need for further research into the mechanisms by which Se levels, selenoprotein expression, and inherited genetic variation in selenoproteins and interacting pathway genes may affect molecular pathways that either prevent or contribute to disease development. This is explored in human cohort studies [1–3,5,6] and in animal [7] and in vitro models [8].

1. Selenium, Cardiovascular Diseases, Metabolic Syndrome, and Obesity

Although links between Se levels, selenoprotein expression, and CVD or lipid metabolism are well established [1,2], the mechanisms supporting these associations remain unclear. Two of the manuscripts presented in this *Issue* established, in distinct Northern European populations, two novel Se-related biomarkers of risk and prognosis for CVD. Schomburg et al. identified, in a Swedish human perspective cohort, a strong correlation between low baseline plasma levels of Selenoprotein P (SELENOP; the major Se-transporter) and risk for all-cause mortality, CVD mortality, and a first CVD event during the follow-up period of 9.3 (8.3–11) years [2]. Based on the known high prevalence (20%) of low SELENOP levels in the Northern European population [2,3], these data suggest that plasma SELENOP concentrations may be used as an early biomarker of CVD risk and, therefore, that targeted Se supplementation could be used as a preventative measure. However, further studies are required to determine the causal link between SELENOP levels and cardiovascular events and to assess whether genetic variants in the *SELENOP* gene, known to influence Se bioavailability, affects this correlation.

Non-selenoprotein Se-binding proteins, such as Selenium-Binding Protein 1 (SELENBP1), have been underappreciated in Se-related research, considering their significant relevance for major physiological processes and that SELENBP1 has been previously implicated in myocardial infarction and poor clinical outcomes from various tumour types [1]. In a distinct cohort of 75 German patients undergoing elective cardiac surgery, a marked but transient increase in circulating SELENBP1 concentrations during the surgical process was correlated, in most patients, with the duration of ischemia and myocardial damage. This suggests that serum concentrations of SELENBP1 could constitute a quantitative marker for myocardial hypoxia [1]. On the contrary, high serum Se levels were associated with metabolic syndrome and insulin resistance markers in a large Taiwanese cohort, although markers of adiposity and lipid functions varied by sex [5].

Further evidence of the importance of Se in lipid and energy metabolism is supported by the observation that the enzyme selenocysteine lyase (Scly) is involved in weight gain, resistance to anorexigenic hormone leptin, and thermogenesis [7]. Scly is responsible for the breakdown of intracellular selenocysteine to alanine and selenide, which is in turn recycled for the synthesis of new selenoproteins. The conditional Scly knockout (KO) model affecting agouti-related peptide-positive neurons in the hypothalamus (Scly-Agrp KO mice) displayed a reduction in high fat diet-induced weight gain and protection against development of leptin resistance compared with controls (mice expressing Scly). This was linked to progressive degeneration of some Agrp neurons and in brown adipose tissue pattern [7].

Taken together, these data provide some novel potential targets for understanding the mechanisms by which Se affects CVD or lipid metabolism and for further investigating the potential of SELENOP and SELENBP1 as CVD biomarkers.

2. Selenium and Cancer

Accumulating experimental and observational evidence suggests that insufficient Se intake and/or selenoprotein genetic variations may contribute to the development of several tumours including colorectal cancer (CRC), mediated by oxidative and inflammatory stress response selenoproteins [3].

Higher Se status levels (total serum Se levels and SELENOP concentrations) were previously reported in a multi-centre, European prospective cohort study (EPIC) to be associated with a decreased CRC risk. The manuscript in this issue by Fedirko et al. extends this work by describing the largest association with CRC risk for common genetic variations related to Se metabolism in approximately 1400 cases and 1400 controls within this EPIC cohort. The study examined over 1000 single nucleotide polymorphisms (SNPs) in 154 genes within the Se biological pathway (including all selenoprotein genes and Se metabolic pathway genes) and interactions with serum Se status biomarkers from the previous study. The findings provide the most comprehensive evidence to date that individual genotypes relevant for selenoprotein expression, metabolism, and function and interaction with Se status may affect CRC risk in a population of marginally low Se status, such as in Europe. Pathway analyses indicated that, for genes in antioxidant/redox and apoptotic pathways, the influence of SNPs on the disease risk is also dependent on interaction with Se status [3].

Addressing the sparse data on selenoprotein expression in CRC, Hughes and colleagues assessed selenoprotein gene transcript levels in the neoplastic and matched mucosal tissue from Irish and Czech colorectal adenoma (CRA) and CRC patients and examined the interaction with Se status levels [6]. Several selenoproteins (including biological stress response and Se biosynthesis genes) were differentially expressed in the disease tissue compared to the normal tissue of both CRA and CRC patients, and that also showed tumour gene expression changes correlated to levels of Se or SELENOP. Across the disease tissues from the adenoma and both cancer groups, *GPX2* and *TXNRD3* exhibited higher expression while *GPX3*, *SELENOP*, *SELENOS*, and *SEPHS2* showed lower expression. The authors concluded that selenoprotein expression changes could be used as biomarkers of functional Se status and the colorectal adenoma to cancer transition. In survival analyses, only a higher *SELENOF* expression was associated with poorer survival outcomes after cancer diagnosis. Although this did not retain significance after multiple testing correction, there is possible biological validity to this observation as *SELENOF* has been previously linked with oncogenesis [6].

Moving to innovative Se metabolism cell-line experiments, Sonet et al. suggest that selenized lipids from plant oils (selenitriglycerides; Selol), proposed to have antineoplastic effects, may provide natural Se supplementation as a bioavailable selenocompound with lower toxicity than chemical forms like selenite [8]. The authors showed that Selol could be an efficient source of Se for selenoprotein biosynthesis in immortalized kidney (HEK293) and prostate cancer (LNCaP) cell lines but not in immortalized prostate cells (PNT1A,) possibly due to variance in lipid metabolism between the different cell lines [8]. As transformation of various chemical species into selenide is the gateway step for further incorporation into selenoproteins, cell-specific Se metabolism to selenide via selenized triglycerides requires further study.

3. Selenium and Responses to Stress

The importance of Se status in modifying the ability of an individual to respond to stress has been linked to the development of many diseases including CVD [1,2], cancers [3,6], and neurodegeneration [9]. As a number of selenoproteins play a crucial prevalent role in the response to oxidative stress and the control of redox status [3,10] as well as in the maintenance of endoplasmic reticulum (ER) stress response [9], mechanisms that alter the expression of these selenoproteins have the potential to lead to an increased risk of disease.

In a currently highly pertinent review, considering the current pandemic of the SARS-CoV-2 virus mediated COVID-19 disease as we write, Guillin et al. discuss the role of Se in protection from viral contagion [10]. During viral infection, the pathogens induce oxidative stress by generating reactive oxygen species and by altering the cellular antioxidant defences, including selenoproteins such as glutathione peroxidases (GPx) and thioredoxin reductases. Consequently, the host's Se requirements increase and, in hosts deficient in Se, the oxidative stress can induce viral genome mutations, leading to increased microbial virulence (e.g., coxsackie and influenza viruses). Other mechanisms by which the host's nutritional status can affect viral infection progression include reducing the ability of the immune system to respond to the virus (e.g., human immunodeficiency virus (HIV) and hepatitis C and B viruses). In silico data have also revealed the presence of selenoprotein genes in the genomes of several common viruses (e.g., HIV) that resemble mammalian GPx. The function and regulation of such viral selenoproteins remains unclear but could afford viruses protection from oxidative damage [10].

Due to the crucial role of ER stress in many cellular processes, understanding the consequences of alteration of components of the ER stress response has the potential to lead to the discovery of new therapeutic/nutritional targets. Seven selenoproteins are known to be present in the ER, but not all have been well characterised. In a timely review, Ren et al. discuss the current knowledge on ER-resident SELENOF, including its function and role in ER stress response and the regulation of its expression [9]. The review also summarises results from genetic association studies linking genotypes for SNPs in the *SELENOF* gene to risk for various cancers, Kashin–Beck disease, and AIDs progression, with a particular focus on two well-characterised functional SNPs (rs5845 and rs5859) affecting SELENOF protein expression. Furthermore, the authors discuss the dysregulation of SELENOF expression in several tissue and pathologies, from cancer to neurodegeneration and immune system diseases [9]. Future studies investigating the role of other ER-resident selenoproteins could lead to a better understanding of mechanisms that contribute to the development of a wide range of common complex diseases.

Overall, the studies in this Special Issue strengthen and broaden the evidence base that the risk of several chronic diseases and viral infections may be modified by Se status, genotype, sex, and gene variation interactions within biological pathways. Detailed investigation of Se intake levels and metabolism is needed to more fully elucidate the relevance for disease etiopathogenesis, especially for populations with diverse Se status levels and/or individuals with potentially at-risk disease or protective Se pathway genotypes.

Author Contributions: Conceptualization, C.M.; Writing–original draft, C.M. and D.J.H.; Writing–review & editing, C.M. and D.J.H. All authors have read and agreed to the published version of the manuscript.

Funding: This research received no external funding.

Conflicts of Interest: The authors declare no conflict of interest. Though, combined, they are co-authors on two of the papers in this special review [3,6], all submissions were considered impartially in this editorial commentary.

Nutrients **2020**, *12*, 1049

References

1. Kühn-Heid, E.C.D.; Kühn, E.; Ney, J.; Wendt, S.; Seelig, J.; Schwiebert, C.; Minich, W.; Stoppe, C.; Schomburg, L. Selenium-Binding Protein 1 Indicates Myocardial Stress and Risk for Adverse Outcome in Cardiac Surgery. *Nutrients* **2019**, *11*, 2005. [CrossRef] [PubMed]
2. Schomburg, L.; Orho-Melander, M.; Struck, J.; Bergmann, A.; Melander, O. Selenoprotein-P Deficiency Predicts Cardiovascular Disease and Death. *Nutrients* **2019**, *11*, 1852. [CrossRef] [PubMed]
3. Fedirko, V.; Jenab, M.; Méplan, C.; Jones, J.; Zhu, W.; Schomburg, L.; Siddiq, A.; Hybsier, S.; Overvad, K.; Tjønneland, A.; et al. Association of Selenoprotein and Selenium Pathway Genotypes with Risk of Colorectal Cancer and Interaction with Selenium Status. *Nutrients* **2019**, *11*, 935. [CrossRef] [PubMed]
4. Ibrahim, S.A.Z.; Kerkadi, A.; Agouni, A. Selenium and Health: An Update on the Situation in the Middle East and North Africa. *Nutrients* **2019**, *11*, 1457. [CrossRef] [PubMed]
5. Lu, C.-W.; Chang, H.-H.; Yang, K.-C.; Chiang, C.-H.; Yao, C.-A.; Huang, K.-C. Gender Differences with Dose-Response Relationship between Serum Selenium Levels and Metabolic Syndrome-A Case-Control Study. *Nutrients* **2019**, *11*, 477. [CrossRef] [PubMed]
6. Hughes, D.J.; Kunicka, T.; Schomburg, L.; Liska, V.; Swan, N.; Soucek, P. Expression of Selenoprotein Genes and Association with Selenium Status in Colorectal Adenoma and Colorectal Cancer. *Nutrients* **2018**, *10*, 1812. [CrossRef] [PubMed]
7. Torres, D.J.; Pitts, M.W.; Hashimoto, A.C.; Berry, M.J. Agrp-Specific Ablation of Scly Protects against Diet-Induced Obesity and Leptin Resistance. *Nutrients* **2019**, *11*, 1693. [CrossRef] [PubMed]
8. Sonet, J.; Mosca, M.; Bierla, K.; Modzelewska, K.; Flis-Borsuk, A.; Suchocki, P.; Ksiazek, I.; Anuszewska, E.; Bulteau, A.-L.; Szpunar, J.; et al. Selenized Plant Oil Is an Efficient Source of Selenium for Selenoprotein Biosynthesis in Human Cell Lines. *Nutrients* **2019**, *11*, 1524. [CrossRef] [PubMed]
9. Ren, B.; Liu, M.; Ni, J.; Tian, J. Role of Selenoprotein F in Protein Folding and Secretion: Potential Involvement in Human Disease. *Nutrients* **2018**, *10*, 1619. [CrossRef] [PubMed]
10. Guillin, O.; Vindry, C.; Ohlmann, T.; Chavatte, L. Selenium, Selenoproteins and Viral Infection. *Nutrients* **2019**, *11*, 2101. [CrossRef] [PubMed]

Article

Selenium-Binding Protein 1 Indicates Myocardial Stress and Risk for Adverse Outcome in Cardiac Surgery

Ellen C. D. Kühn-Heid [1], Eike C. Kühn [1], Julia Ney [2], Sebastian Wendt [2,3], Julian Seelig [1], Christian Schwiebert [1], Waldemar B. Minich [1], Christian Stoppe [2] and Lutz Schomburg [1,*]

[1] Institut für Experimentelle Endokrinologie, Charité – Universitätsmedizin Berlin, corporate member of Freie Universität Berlin, Humboldt-Universität zu Berlin, and Berlin Institute of Health, D-13353 Berlin, Germany

[2] Cardiovascular Critical Care & Anesthesia Research and Evaluation (3CARE), RWTH-Aachen University, D-52074 Aachen, Germany

[3] Department of Anesthesiology, Uniklinik RWTH-Aachen, D-52074 Aachen, Germany

* Correspondence: lutz.schomburg@charite.de; Tel.: +49-30-450-524-289

Received: 31 July 2019; Accepted: 22 August 2019; Published: 25 August 2019

Abstract: Selenium-binding protein 1 (SELENBP1) is an intracellular protein that has been detected in the circulation in response to myocardial infarction. Hypoxia and cardiac surgery affect selenoprotein expression and selenium (Se) status. For this reason, we decided to analyze circulating SELENBP1 concentrations in patients ($n = 75$) necessitating cardioplegia and a cardiopulmonary bypass (CPB) during the course of the cardiac surgery. Serum samples were collected at seven time-points spanning the full surgical process. SELENBP1 was quantified by a highly sensitive newly developed immunological assay. Serum concentrations of SELENBP1 increased markedly during the intervention and showed a positive association with the duration of ischemia ($\varrho = 0.6$, $p < 0.0001$). Elevated serum SELENBP1 concentrations at 1 h after arrival at the intensive care unit (post-surgery) were predictive to identify patients at risk of adverse outcome (death, bradycardia or cerebral ischemia, "endpoint 1"; OR 29.9, CI 3.3–268.8, $p = 0.00027$). Circulating SELENBP1 during intervention (2 min after reperfusion or 15 min after weaning from the CPB) correlated positively with an established marker of myocardial infarction (CK-MB) measured after the intervention (each with $\varrho = 0.5$, $p < 0.0001$). We concluded that serum concentrations of SELENBP1 were strongly associated with cardiac arrest and the duration of myocardial ischemia already early during surgery, thereby constituting a novel and promising quantitative marker for myocardial hypoxia, with a high potential to improve diagnostics and prediction in combination with the established clinical parameters.

Keywords: trace element; biomarker; selenoprotein; metabolism; redox regulation; prediction

1. Introduction

Selenium (Se) is an essential trace element affecting the expression of selenoproteins and redox signaling [1,2]. The metabolism of Se and the control of selenoprotein expression is complex as there are disease-, genotype-, sex- and potentially age-specific differences in combination with a variety of nutritional Se sources [3–7]. There are two major types of proteins related to Se, i.e., the selenoproteins with one or more genetically encoded selenocysteine residues in their primary sequence [8] versus the group of less-well-defined Se-binding proteins [9]. Selenium-Binding Protein 1 (SELENBP1) constitutes the most-intensively characterized member of the latter group of proteins [10]. SELENBP1 is expressed in most human and rodent tissues and migrates as a 56 kD band in electrophoretic analyses [11]. Its expression levels have been associated with tumorigenesis and cancer growth [12], especially in relation to androgen concentrations and prostate cancer [13]. A physical interaction between SELENBP1 and the major intracellular Se-dependent glutathione peroxidase (GPX1) has been described, functionally connecting both groups of Se-containing proteins [14]. Several analyses have quantified SELENBP1 in tumor tissues and report on reduced SELENBP1 expression levels in association with poor survival in gastric [15], breast [16] and renal cancer [17]. Collectively, these studies have highlighted the Se-dependent intracellular activities of SELENBP1 and a prognostic value of its expression levels in tumors as a biomarker of disease severity.

Alterations in Se blood levels and differences in the concentration of extracellular selenoproteins appear to be of similar relevance for health and disease. There are three major Se-containing circulating proteins, i.e., the plasma glutathione peroxidase (GPX3), the Se transport protein selenoprotein P (SELENOP) and selenomethionine-containing albumin. Under normal conditions, SELENOP constitutes the major fraction of extracellular Se [4], whereas the contribution of selenomethionine-containing albumin to total Se concentration in blood largely depends on the dietary intake of selenomethionine [18]. Circulating GPX3 is mainly derived from the kidneys and is regulated by hypoxia, oxidative stress and Se status [19]. Genetic inactivation of the murine *Gpx3* gene has not resulted in a major phenotype, except for effects on Se status [20]. The inactivation of *Selenop* in mice gave rise to a complex phenotype with male infertility, growth defects, oxidative stress and neurological impairment [21,22]. In patients, low SELENOP concentrations have been observed in sepsis [23], inflammatory bowel disease [24] or steatohepatitis [25] and shown to correlate to poor survival in renal cancer [26], sepsis [27] or after major trauma [28].

Information on the role and regulation of extracellular SELENBP1 is sparse. A recent report identified the protein in urine as a novel and early biomarker of acute kidney injury [29] and we described increased SELENBP1 concentrations in patients with acute coronary syndrome at high risk of major cardiac events [30]. Cardiac surgery negatively affects Se status and Se deficiency increases the risk of ischaemic heart disease [31] and promotes organ dysfunction after cardiac surgery [32]. To better characterize circulating SELENBP1 as a biomarker of myocardial stress, we monitored its concentration in patients undergoing surgery with cardioplegia-induced myocardial arrest and the use of a cardio-pulmonary bypass (CPB) and evaluated its potential diagnostic value with respect to convalescence and survival.

2. Materials and Methods

2.1. Patients

In this study, consecutive patients scheduled for elective cardiac surgery with a necessity for the use of a cardiopulmonary bypass (CPB) and cardioplegia-induced myocardial arrest were invited to participate in this analytical study. The protocol was approved by the local institutional review board (Ethics committee, RWTH Aachen University, Germany), registered at ClinicalTrials.gov (ClinicalTrials.gov identifier: NCT0126772), and all participants provided informed written consent. Adult patients scheduled for elective cardiac surgery were included, and patients who were unable to give informed consent, patients with suspicious or proven pregnancy or malignancy, and patients with perioperative infections were excluded.

2.2. Clinical Examination, Sample Collection and Analysis

Relevant clinical data were recorded as part of the clinical routine. Serum samples were taken from the central venous line after the induction of anaesthesia (pre-operative) 45 min after the institution of CPB (myocardial ischemia), 2 min after opening of the cross-clamps (myocardial reperfusion), 15 min after weaning from the CPB, as well as 1 h, 6 h and 24 h after admission to the ICU. Samples were stored at $-80\,°C$ until analysis. Several routine parameters were analyzed by the clinical laboratory at Uniklinik RWTH Aachen. SELENBP1 was measured as described earlier [30] at Institut für Experimentelle Endokriologie, Charité - Universitätsmedizin Berlin. Intra- and inter-assay coefficients of variation of SELENBP1 were below 10% each.

2.3. Statistical Analysis

All statistical analyses were performed using freely available statistical software (R 3.5.1, The R Foundation for Statistical Computing). Normal distribution of data was assessed by sample size or visual inspection of the Q-Q plot. Welch's t-test was applied for discrete data if normality could be assumed. Wilcoxon's rank-sum test was used if data was sparse or not normally distributed. When comparing groups of categorical data, Fisher's exact test was applied or Chi-squared test was used for sufficiently large fields of data. Spearman's rank correlation test was performed for testing for correlations. Friedman's test was used when testing for changes over time due to the data not being normally distributed. Error bars for medians could not be shown, as the test assumes normality. In a post-hoc analysis, all the time-points were compared to baseline using Wilcoxon's signed-rank test. Confidence intervals (CI) and significance were reported for an interval of 0.95 and accordingly an α of 0.05, if not stated differently. In the graphics and tables, the star symbols denote the level of significance: *: $p < 0.05$, **: $p < 0.01$, ***: $p < 0.001$, and ****: $p < 0.0001$. The following algorithm was used to determine quantiles: m = (p + 1)/3. p[k] = (k-1/3)/(n + 1/3), resulting in p[k] = median[F(x[k])]. Cross-tables and empirical receiver-operator-curves (ROC) were used to establish threshold concentrations above which the risk for adverse outcomes increased. If a threshold was found to be of diagnostic or predictive value, the area under the curve (AUC) and p-value were calculated and reported.

3. Results

Baseline characteristics of the patients enrolled in the clinical study are provided below. The patients underwent either a single procedure or a procedure combining two or more interventions (Table 1).

Table 1. Clinical characteristics of patients in relation to serum selenium-binding protein (1SELENBP1) concentrations.

Parameter	Total (*n* = 66)	SELENBP1-Positive * (*n* = 18)	SELENBP1-Negative (*n* = 48)
Age (years)	65 (58–75)	65 (60–75)	66 (56–75)
Female sex (*n*)	23% (15)	8% (5)	15% (10)
Male sex (n)	77% (51)	20% (13)	58% (38)
BMI (kg/m^2)	27 (25–31)	27 (25–32)	28 (25–30)
EUROscore [1]	4 (2–6)	6 (5–8)	3 (1–6)
Duration of intervention (min)	271 (207–330)	282 (225–317)	265 (200–331)
Duration of CPB (min)	120 (85–149)	148 (126–176)	100 (83–134)
Duration of ischemia (min)	72 (58–107)	112 (93–134)	63 (58–84)
Stay in hospital (days)	9 (7–11)	11 (8–14)	9 (7–10)
Bradycardia (n)	6% (4)	5% (3)	2% (1)
Exitus letalis (n)	5% (3)	5% (3)	0% (0)
Pneumonia (n)	5% (3)	5% (3)	0% (0)
Acute kidney failure (n)	5% (3)	3% (2)	2% (1)
Wound infection (n)	2% (1)	2% (1)	0% (0)
Cerebral ischemia (n)	2% (1)	2% (1)	0% (0)

Data are expressed as % (n) or median (IQR); * SELENBP1 positive; serum SELENBP1 concentration > 0.96 nM at 1 h after arrival on the ICU; [1] EURO score; system to assess early mortality after cardiac surgery, according to Nashed et al. [33].

3.1. Preoperative Status of SELENBP1 Follows a Normal Distribution

The patients analyzed in this study underwent elective cardiac surgery, as opposed to an acute emergency procedure. For this reason, it was assumed that baseline serum SELENBP1 concentrations would follow a normal distribution before the intervention. Indeed, SELENBP1 values were normally distributed with the 99th percentile at 0.4 nM and the mean + 2.5 standard deviations at 0.3 nM (Figure 1), similarly to the values reported from a healthy cohort [30]. Baseline SELENBP1 concentrations were not related to perioperative risk as assessed with the EUROscore (Table 1). Two outliers were identified who started with markedly elevated baseline concentrations of circulating SELENBP1 (z = 178.2 and z = 23.6, respectively).

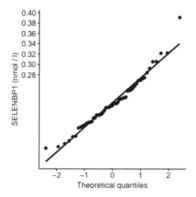

Figure 1. QQ-Plot of SELENBP1 in serum at baseline. The plot indicates the distribution of serum SELENBP1 concentrations at baseline before surgical intervention. The straight line indicates a normal distribution (R^2 = 0.96, $p < 0.0001$).

3.2. Intraoperative and Postoperative Kinetics of SELENBP1 Indicate a Transient Increase

The blood samples taken at different time points during and after the surgical procedure were analyzed in order to assess the kinetics of circulating SELENBP1. The serum concentrations significantly increased as compared to baseline over the full course of the treatment (Figure 2).

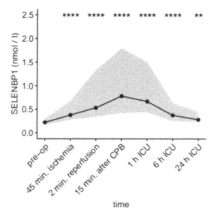

Figure 2. SELENBP1 before, during and after surgery. Median and 10th–90th percentile of serum SELENBP1 concentrations are provided. CPB: cardio-pulmonary bypass, ICU: intensive care unit. The values were compared to baseline by Wilcoxon's signed-rank test. **; $p = 0.0072$, ****; $p < 0.0001$.

During intervention, serum concentrations of SELENBP1 increased by ≥20% in n = 72 of the cardiac patients (96%, Figure 3A), and 68 patients (91%, Figure 3B) surpassed the cut-off at 0.46 nM, determined earlier as the threshold for an increased risk of death or other major adverse cardiac event in patients with suspected myocardial infarction [30]. The SELENBP1 concentrations increased until 2 min after reperfusion in 61 patients (81%), and until 15 min after weaning from the CPB in 71 patients (95%). Notably, the increase was transient, highlighting an inducing stimulus of short duration. Declining serum concentrations of SELENBP1 were observed in 65 patients (87%) within the first 6 h after admission to the ICU and in 70 patients (93%) within 24 h after admission to the ICU. In total, 74 patients (99%) showed a transient increase in serum SELENBP1 concentrations after reperfusion and qualified as responders in our hypothesis.

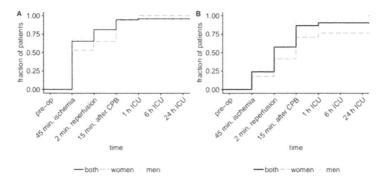

Figure 3. Transient SELENBP1 increased above thresholds. (**A**) Almost all the patients displayed elevated serum SELENBP1 concentrations during the intervention, as judged by an increase of at least 20% in comparison to pre-op levels, or (**B**) by surpassing a predefined cut-off value (0.455 nM), or an increase by ≥20% if the pre-op level was already elevated and >0.455 nM (empirical cumulative distribution).

3.3. Serum SELENBP1 Concentrations Correlate to the Duration of Ischemia

Cardioplegia and the CPB exert a severe stress on the myocardium and the whole organism. Both interventions are related to ischemia of the heart and may constitute the major stimulus for the release of SELENBP1 into the bloodstream. Accordingly, a strong association of serum SELENBP1 concentrations with the duration of ischemia was observed already at 2 min after reperfusion ($\varrho = 0.5$, $p < 0.0001$), and at 15 min after weaning from the CPB ($\varrho = 0.6$, $p < 0.0001$, Figure 4). Serum SELENBP1 was slightly lower in female than in male patients at 15 min after CPB (median for female sex: 0.6 nM vs. male sex: 0.8 nM, $p = 0.043$). Fewer female patients showed a rise in SELENBP1 concentrations until 2 min after reperfusion (0.6 of female patients vs. 0.9 of male patients, $p = 0.045$), indicating a slightly lower velocity. The peak values did not differ significantly between the sexes.

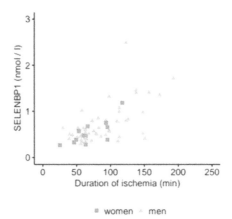

Figure 4. Correlation of serum SELENBP1 concentrations with duration of ischemia. Serum SELENBP1 concentrations were determined at 15 min after weaning from the CBP and are displayed in relation to duration of ischemia. A linear correlation is observed ($\varrho = 0.6$, $p < 0.0001$).

3.4. Elevated Serum SELENBP1 Concentrations Predict Myocardial Damage

Circulating concentrations of SELENBP1 were unrelated to many routine laboratory markers in patients attending the emergency ward for chest pain, like troponin T, creatinine, the heart-specific isoform of creatine kinase (CK-MB), the liver enzymes ASAT or ALAT, white blood cell count (WBC), potassium or others [30]. In contrast to the former study, this analysis used samples where the time points for analysis were controlled tightly and were pre-defined, allowing a close monitoring and comparison of the kinetic behavior of cardiac biomarkers. Serum SELENBP1 concentrations increased already intraoperatively (2 min after reperfusion or 15 min after weaning from the CPB) and correlated positively to the CK-MB values measured after the intervention. This result indicates that a rise in serum SELENBP1 concentrations constituted an early event and preceded a rise in CK-MB concentrations as a marker for myocardial damage (Figure 5).

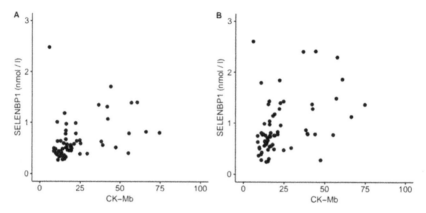

Figure 5. Correlation of intraoperative serum SELENBP1 concentrations with heart-specific creatinine kinase (CK-MB) after surgery. Serum SELENBP1 measured during intervention and heart-specific CK-MB concentrations measured after intervention display a positive correlation, both to the SELENBP1 concentrations measured at (**A**) 2 min after reperfusion ($\varrho = 0.5$, $p < 0.0001$), and at (**B**) 15 min after weaning from the CBP ($\varrho = 0.4$, $p = 0.0007$), respectively.

3.5. Elevated Serum SELENBP1 Concentrations Are Indicative of Adverse Outcomes

Next, the potential association of serum SELENBP1 with adverse outcomes was evaluated. The elevated serum concentrations of SELENBP1 were significantly associated with adverse outcomes, and predictive to discriminate patients at risk from those with negligible risk for death, bradycardia, cerebral ischemia, stay at the ICU for > 40 days, acute kidney injury or pneumonia (Table 2).

Patients suffering from bradycardia, death, or cerebral ischemia (combined endpoint 1) displayed markedly elevated serum concentrations of SELENBP1 at several points compared to those patients without any of these serious events ($p = 0.033$ at 45 min after induction of ischemia, $p = 0.0051$ at 2 min after reperfusion, $p = 0.0011$ at 15 min after weaning from the CPB, and $p = 0.0017$ at 1 h after arrival on the ICU) (Figure 6).

Table 2. Association of elevated serum SELENBP1 concentrations with adverse outcomes.

Parameter	Threshold [nM]	OR	CI	p	PPV	NPV	AUC	n
15 min after CPB								
combined endpoint 1 *	0.988	22	3–194	0.0005	0.333	0.978	0.84	8
combined endpoint 2 *	0.820	18	2–152	0.0011	0.310	0.976	0.81	9
1 h ICU								
death	2.238	124	6.2777	0.0042	0.667	0.984	0.92	2
acute kidney injury	0.889	20	1–411	0.029	0.158	1.000	0.83	4
combined endpoint 1 *	0.956	30	3–269	0.0003	0.389	0.979	0.84	7
combined endpoint 2 *	0.889	33	4–296	<0.0001	0.421	0.979	0.84	7
6 h ICU								
death	1.292	63	4–1018	0.0077	0.500	0.984	0.88	3
maximal levels								
bradycardia	0.988	23	1–439	0.014	0.167	1.000	0.81	5
cerebral ischemia	1.396	25	1–544	0.034	0.143	1.000	0.90	3
combined endpoint 1 *	0.988	25	3–215	0.0003	0.333	0.980	0.88	8
combined endpoint 2 *	0.968	24	3–199	0.0003	0.333	0.979	0.86	9
stay in ICU for >40 day	0.870	4	1–10	0.013	0.484	0.795	0.70	14

* Combined endpoints; combined endpoint 1 comprises bradycardia, death, and cerebral ischemia; combined endpoint 2 comprises bradycardia, death, cerebral ischemia, as well as acute kidney injury. OR; odds ratio; CI, confidence interval; PPV, positive predictive value; NPV, negative predictive value, AUC, area under the curve.

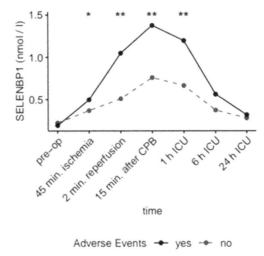

Adverse Events → yes → no

Figure 6. SELENBP1 before, during and after cardiac surgery in relation to outcome. Elevated serum SELENBP1 concentrations are associated with adverse events during the full course of surgery, especially directly after reperfusion, after weaning from the CPB, and shortly after arriving on the ICU. Adverse events comprise death, cerebral ischemia, bradycardia ($p = 0.033$ for 45 min after induction of ischemia, $p = 0.0051$ for 2 min after reperfusion, $p = 0.0011$ for 15 min after weaning from the CPB, and $p = 0.0017$ for 1 h after arrival on the ICU) (AUC; 0.8). **; $p < 0.01$, *; $p < 0.05$.

The clinical data along with the information on serum SELENBP1 were finally used to conduct a Receiver-Operator-Curve (ROC) analysis for predicting adverse events on the basis of elevated serum concentrations of SELENBP1 (Figure 7).

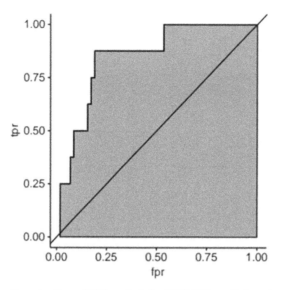

Figure 7. Receiver-Operator-Curve (ROC) analysis for SELENBP1 predicting adverse events. The receiver-operator-analysis for SELENBP1 concentrations and adverse events (death, cerebral ischemia, bradycardia) indicates a relatively meaningful predictive value (AUC; 0.8).

4. Discussion

In this study, circulating SELENBP1 concentrations were monitored in patients undergoing cardiac surgery along the time span from preparation for the intervention until 24 h after arriving at the ICU. The surgical intervention involved a CPB, and this procedure constitutes a clinical model for ischemia and reperfusion. Importantly, all time points could be reliably analyzed as the patients were not arriving at the hospital after an adverse incidence but underwent an elective and tightly scheduled procedure. To our surprise, a strong increase in serum SELENBP1 concentrations was detected in all but one patient. The SELENBP1 concentrations in serum reflected the myocardial damage already during surgery and nicely correlated with the levels of the heart-specific isoform of creatine kinase (CK-MB) measured after the intervention. The comparison of serum SELENBP1 concentrations with the duration of ischemia indicated an almost linear relationship, suggesting that serum SELENBP1 constituted a reliable novel marker of myocardial stress.

This notion is supported by the strong association of circulating SELENBP1 levels during and directly after the procedure, with a higher mortality risk and incidence rate of adverse events, including, e.g., length of stay at the ICU, bradycardia or cerebral ischemia. The thresholds with good positive and strong negative predictive power were in the same range as determined before in relation to chest pain and suspected myocardial infarction [30]. Collectively, serum concentrations of SELENBP1 seem to reliably indicate severe cardiac damage in patients, whereas no relevant amounts of serum SELENBP1 were detectable under control conditions in healthy subjects where SELENBP1 is located intracellularly [10,14,34]. Notably, the increased levels and fast kinetics of circulating SELENBP1 were not affected by traditional risk-factors and the baseline values did not reflect perioperative risk as assessed by the EURO-Score [33], supporting the notion that circulating SELENBP1 reflects acute events rather than indicating chronic risk-related factors. It remains to be studied whether the extent of SELENBP1 release is related to baseline Se status, as determined by the activity of circulating GPX3, SELENOP or total serum Se concentrations, and whether SELENBP1 is increased in chronic inflammatory heart disease as a potentially valuable prospective biomarker.

The previous study indicated that elevated serum concentrations of SELENBP1 in patients suspected of myocardial infarction were indicative of a higher mortality and other major adverse cardiac events [30]. There were no significant correlations with the commonly used clinical markers, such as troponin, as measured with a high-sensitivity assay, aspartate aminotransferase, creatinine, fibrinogen, prothrombin time or blood cell counts [30]. Consequently, we assumed that the stimulus for the release of SELENBP1 into the bloodstream should be different from necrosis of cardiomyocytes, which would be associated with increased circulating cardial troponins. The fast appearance of SELENBP1 already during surgery and the linear relationship between circulating SELENBP1 and the duration of myocardial ischemia suggests that SELENBP1 may have been released by cardiomyocytes upon stress and before necrosis. This hypothesis is supported by the moderate linear correlation of serum SELENBP1 during surgery to CK-MB measured at later time-points and the biologically plausible time span needed for necrosis to occur in response to hypoxia and acidosis [35].

Heart tissue is not the only potential source of extracellular SELENBP1, as urinary SELENBP1 was described as a new biomarker of heavy metal-induced kidney injury, where the protein increased strongly in the renal cortex after mercury administration in a rodent model of nephrotoxicity [36]. The authors propose urinary SELENBP1 as a novel sensitive and specific biomarker for early stages of kidney injury before irreversible damage has taken place [29,36]. Accordingly, serum SELENBP1 may indicate severe and acute noxae to cardiomyocytes at an early time point before the onset of necrosis, i.e., at a time when protective measures may possibly reverse the path to cell death. In this context, it may be speculated that a secretion of SELENBP1 caused an associated decline of intracellular Se stores, thereby depriving the cardiomyocyte of the trace element with essential importance for antioxidative defense systems [37]. This notion is supported by clinical studies, where supplementation with Se along with coenzyme Q10 significantly reduced cardiovascular mortality in a prospective randomized double-blind controlled intervention study in European seniors [38].

There was an unexpected small difference between male and female patients concerning the velocity of the increase. However, peak values were similar and larger studies are needed in order to elucidate whether SELENBP1 secretion into the circulation constitutes another example of sex-specific differences in Se metabolism and selenoprotein expression [5]. The initial symptoms, disease development and mortality risk upon myocardial infarction show strong sex-specific differences that attenuate with age, highlighting sex as an important modifier of cardiovascular physiology and heart disease course [39]. It will therefore be important to conduct larger studies with both male and female patients in order to elucidate the importance of sex and age for the increase of circulating SELENBP1 concentrations, its dynamics and predictive value.

A limitation of this clinical study constitutes the relatively small sample size, limiting the results to testing the major hypothesis without providing the power for detailed and stratified analyses of subgroups of patients or sex-specific differences. A major strength of this study was the reliable, sensitive and highly reproducible analytical technique used, i.e., the novel luminometric immunoassay used for SELENBP1 quantification, which is based on specific monoclonal antibodies and is therefore available for verification analyses and additional clinical studies.

5. Conclusions

In this study, the kinetics and modifying factors of circulating SELENBP1 in response to myocardial injury were characterized. Increased concentrations reflected the duration of ischemia and myocardial damage, and reliably identified patients at risk of adverse outcomes. It is hoped that these insights will be useful for improving the care of cardiac surgery patients and help to identifying individuals with high risks for adverse events in order to raise increased attention and enable fast curative measures when first signs of worsening of the clinical condition become apparent.

Author Contributions: E.C.D.K.-H., E.C.K., C.S. (Christian Stoppe) and L.S. designed the study. E.C.D.K.-H. and E.C.K. did the statistical analyses. E.C.D.K.-H., E.C.K., C.S. (Christian Stoppe) and LS wrote the manuscript. E.C.D.K.-H., E.C.K., J.N., S.W., J.S., C.S. (Christian Schwiebert), W.B.M. and C.S. (Christian Stoppe) collected the material and conducted the measurements, respectively. All authors critically reviewed the final manuscript and approved its submission.

Funding: Schomburg and Stoppe were supported by grants from Deutsche Forschungsgemeinschaft (DFG Research Unit 2558 TraceAge, Scho 849/6-1 to LS; and STO 1099/-2 to CS). In addition, we acknowledge support from the DFG and the Open Access Publication Fund of Charité – Universitätsmedizin Berlin.

Acknowledgments: We thank Vartitér Seher and Katja Schreiber for excellent technical support, and Andreas Goetzenich, RWTH Aachen, for helpful advice during the initiation of this study.

Conflicts of Interest: J.N., S.W., J.S., C.S. (Christian Schwiebert) and C.S. (Christian Stoppe) have nothing to declare in relation to this study. L.S. holds shares in selenOmed GmbH, a company involved in Se status assessment and supplementation. E.C.D.K.-H., E.C.K., W.B.M. and L.S. are named as inventors on a related patent application.

References

1. Hondal, R.J.; Marino, S.M.; Gladyshev, V.N. Selenocysteine in thiol/disulfide-like exchange reactions. *Antioxid. Redox Signal.* **2013**, *18*, 1675–1689. [CrossRef] [PubMed]
2. Brigelius-Flohé, R.; Flohé, L. Selenium and redox signaling. *Arch. Biochem. Biophys.* **2017**, *617*, 48–59. [CrossRef] [PubMed]
3. Méplan, C. Selenium and chronic diseases: A nutritional genomics perspective. *Nutrients* **2015**, *7*, 3621–3651. [CrossRef] [PubMed]
4. Burk, R.F.; Hill, K.E. Regulation of Selenium Metabolism and Transport. *Ann. Rev. Nutr.* **2015**, *35*, 109–134. [CrossRef] [PubMed]
5. Schomburg, L.; Schweizer, U. Hierarchical regulation of selenoprotein expression and sex-specific effects of selenium. *Biochim. Biophys. Acta* **2009**, *1790*, 1453–1462. [CrossRef] [PubMed]
6. Schomburg, L.; Riese, C.; Renko, K.; Schweizer, U. Effect of age on sexually dimorphic selenoprotein expression in mice. *Biol. Chem.* **2007**, *388*, 1035–1041. [CrossRef] [PubMed]
7. Rayman, M.P. Food-chain selenium and human health: Emphasis on intake. *Br. J. Nutr.* **2008**, *100*, 254–268. [CrossRef]
8. Labunskyy, V.M.; Hatfield, D.L.; Gladyshev, V.N. Selenoproteins: Molecular pathways and physiological roles. *Physiol. Rev.* **2014**, *94*, 739–777. [CrossRef]
9. Bansal, M.P.; Oborn, C.J.; Danielson, K.G.; Medina, D. Evidence for two selenium-binding proteins distinct from glutathione peroxidase in mouse liver. *Carcinogenesis* **1989**, *10*, 541–546. [CrossRef]
10. Elhodaky, M.; Diamond, A.M. Selenium-Binding Protein 1 in Human Health and Disease. *Int. J. Mol. Sci.* **2018**, *19*, 3437. [CrossRef]
11. Chang, P.W.; Tsui, S.K.; Liew, C.; Lee, C.C.; Waye, M.M.; Fung, K.P. Isolation, characterization, and chromosomal mapping of a novel cDNA clone encoding human selenium binding protein. *J. Cell Biochem.* **1997**, *64*, 217–224. [CrossRef]
12. Lanfear, J.; Fleming, J.; Walker, M.; Harrison, P. Different patterns of regulation of the genes encoding the closely related 56 kDa selenium- and acetaminophen-binding proteins in normal tissues and during carcinogenesis. *Carcinogenesis* **1993**, *14*, 335–340. [CrossRef] [PubMed]
13. Yang, M.; Sytkowski, A.J. Differential expression and androgen regulation of the human selenium-binding protein gene hSP56 in prostate cancer cells. *Cancer Res.* **1998**, *58*, 3150–3153. [PubMed]
14. Fang, W.; Goldberg, M.L.; Pohl, N.M.; Bi, X.; Tong, C.; Xiong, B.; Koh, T.J.; Diamond, A.M.; Yang, W. Functional and physical interaction between the selenium-binding protein 1 (SBP1) and the glutathione peroxidase 1 selenoprotein. *Carcinogenesis* **2010**, *31*, 1360–1366. [CrossRef] [PubMed]
15. Xia, Y.J.; Ma, Y.Y.; He, X.J.; Wang, H.J.; Ye, Z.Y.; Tao, H.Q. Suppression of selenium-binding protein 1 in gastric cancer is associated with poor survival. *Hum. Pathol.* **2011**, *42*, 1620–1628. [CrossRef] [PubMed]
16. Zhang, S.; Li, F.; Younes, M.; Liu, H.; Chen, C.; Yao, Q. Reduced selenium-binding protein 1 in breast cancer correlates with poor survival and resistance to the anti-proliferative effects of selenium. *PLoS ONE* **2013**, *8*, e63702. [CrossRef] [PubMed]
17. Ha, Y.S.; Lee, G.T.; Kim, Y.H.; Kwon, S.Y.; Choi, S.H.; Kim, T.H.; Kwon, T.G.; Yun, S.J.; Kim, I.Y.; Kim, W.J. Decreased selenium-binding protein 1 mRNA expression is associated with poor prognosis in renal cell carcinoma. *World J. Surg. Oncol.* **2014**, *12*, 288. [CrossRef] [PubMed]
18. Burk, R.F.; Hill, K.E.; Motley, A.K. Plasma selenium in specific and non-specific forms. *BioFactors* **2001**, *14*, 107–114. [CrossRef]

19. Bierl, C.; Voetsch, B.; Jin, R.C.; Handy, D.E.; Loscalzo, J. Determinants of human plasma glutathione peroxidase (GPx-3) expression. *J. Biol. Chem.* **2004**, *279*, 26839–26845. [CrossRef]
20. Burk, R.F.; Olson, G.E.; Winfrey, V.P.; Hill, K.E.; Yin, D. Glutathione peroxidase-3 produced by the kidney binds to a population of basement membranes in the gastrointestinal tract and in other tissues. *Am. J. Physiol. Gastrointest. Liver Physiol.* **2011**, *301*, G32–G38. [CrossRef]
21. Schomburg, L.; Schweizer, U.; Holtmann, B.; Flohé, L.; Sendtner, M.; Köhrle, J. Gene disruption discloses role of selenoprotein P in selenium delivery to target tissues. *Biochem. J.* **2003**, *370*, 397–402. [CrossRef] [PubMed]
22. Hill, K.E.; Zhou, J.; McMahan, W.J.; Motley, A.K.; Atkins, J.F.; Gesteland, R.F.; Burk, R.F. Deletion of selenoprotein P alters distribution of selenium in the mouse. *J. Biol. Chem.* **2003**, *278*, 13640–13646. [CrossRef] [PubMed]
23. Hollenbach, B.; Morgenthaler, N.G.; Struck, J.; Alonso, C.; Bergmann, A.; Köhrle, J.; Schomburg, L. New assay for the measurement of selenoprotein P as a sepsis biomarker from serum. *J. Trace Elem. Med. Biol. Organ Soc. Miner. Trace Elem. (GMS)* **2008**, *22*, 24–32. [CrossRef] [PubMed]
24. Barrett, C.W.; Short, S.P.; Williams, C.S. Selenoproteins and oxidative stress-induced inflammatory tumorigenesis in the gut. *Cell. Mol. Life Sci. CMLS* **2017**, *74*, 607–616. [CrossRef] [PubMed]
25. Polyzos, S.A.; Kountouras, J.; Mavrouli, M.; Katsinelos, P.; Doulberis, M.; Gavana, E.; Duntas, L. Selenoprotein P in Patients with Nonalcoholic Fatty Liver Disease. *Exp. Clin. Endocrinol. Diabetes Off. J. Ger. Soc. Endocrinol. Ger. Diabetes Assoc.* **2019**. [CrossRef] [PubMed]
26. Meyer, H.A.; Endermann, T.; Stephan, C.; Stoedter, M.; Behrends, T.; Wolff, I.; Jung, K.; Schomburg, L. Selenoprotein P status correlates to cancer-specific mortality in renal cancer patients. *PLoS ONE* **2012**, *7*, e46644. [CrossRef] [PubMed]
27. Forceville, X.; Mostert, V.; Pierantoni, A.; Vitoux, D.; Le Toumelin, P.; Plouvier, E.; Dehoux, M.; Thuillier, F.; Combes, A. Selenoprotein P, rather than glutathione peroxidase, as a potential marker of septic shock and related syndromes. *Eur. Surg. Res. Eur. Chir. Forsch. Rech. Chir. Eur.* **2009**, *43*, 338–347. [CrossRef]
28. Braunstein, M.; Kusmenkov, T.; Zuck, C.; Angstwurm, M.; Becker, N.P.; Bocker, W.; Schomburg, L.; Bogner-Flatz, V. Selenium and Selenoprotein P Deficiency Correlates with Complications and Adverse Outcome After Major Trauma. *Shock (AugustaGa)* **2019**. [CrossRef]
29. Kim, K.S.; Yang, H.Y.; Song, H.; Kang, Y.R.; Kwon, J.; An, J.; Son, J.Y.; Kwack, S.J.; Kim, Y.M.; Bae, O.N.; et al. Identification of a sensitive urinary biomarker, selenium-binding protein 1, for early detection of acute kidney injury. *J. Toxicol. Environ. Health Part A* **2017**, *80*, 453–464. [CrossRef]
30. Kühn, E.C.; Slagman, A.; Kühn-Heid, E.C.D.; Seelig, J.; Schwiebert, C.; Minich, W.B.; Stoppe, C.; Möckel, M.; Schomburg, L. Circulating levels of selenium-binding protein 1 (SELENBP1) are associated with risk for major adverse cardiac events and death. *J. Trace Elem. Med. Biol. Organ Soc. Miner. Trace Elem. (GMS)* **2019**, *52*, 247–253. [CrossRef]
31. Suadicani, P.; Hein, H.O.; Gyntelberg, F. Serum selenium concentration and risk of ischaemic heart disease in a prospective cohort study of 3000 males. *Atherosclerosis* **1992**, *96*, 33–42. [CrossRef]
32. Wendt, S.; Schomburg, L.; Manzanares, W.; Stoppe, C. Selenium in Cardiac Surgery. *Nutr. Clin. Pract. Off. Publ. Am. Soc. Parenter. Enter. Nutr.* **2019**, *34*, 528–539. [CrossRef] [PubMed]
33. Nashef, S.A.; Roques, F.; Michel, P.; Gauducheau, E.; Lemeshow, S.; Salamon, R. European system for cardiac operative risk evaluation (EuroSCORE). *Eur. J. Cardio Thorac. Surg.* **1999**, *16*, 9–13. [CrossRef]
34. Ying, Q.; Ansong, E.; Diamond, A.M.; Yang, W. A Critical Role for Cysteine 57 in the Biological Functions of Selenium Binding Protein-1. *Int. J. Mol. Sci.* **2015**, *16*, 27599–27608. [CrossRef] [PubMed]
35. Graham, R.M.; Frazier, D.P.; Thompson, J.W.; Haliko, S.; Li, H.; Wasserlauf, B.J.; Spiga, M.G.; Bishopric, N.H.; Webster, K.A. A unique pathway of cardiac myocyte death caused by hypoxia-acidosis. *J. Exp. Biol.* **2004**, *207*, 3189–3200. [CrossRef] [PubMed]
36. Lee, E.K.; Shin, Y.J.; Park, E.Y.; Kim, N.D.; Moon, A.; Kwack, S.J.; Son, J.Y.; Kacew, S.; Lee, B.M.; Bae, O.N.; et al. Selenium-binding protein 1: A sensitive urinary biomarker to detect heavy metal-induced nephrotoxicity. *Arch. Toxicol.* **2017**, *91*, 1635–1648. [CrossRef] [PubMed]
37. Boucher, F.R.; Jouan, M.G.; Moro, C.; Rakotovao, A.N.; Tanguy, S.; de Leiris, J. Does selenium exert cardioprotective effects against oxidative stress in myocardial ischemia? *Acta Physiol. Hung.* **2008**, *95*, 187–194. [CrossRef]

38. Alehagen, U.; Aaseth, J.; Johansson, P. Reduced Cardiovascular Mortality 10 Years after Supplementation with Selenium and Coenzyme Q10 for Four Years: Follow-Up Results of a Prospective Randomized Double-Blind Placebo-Controlled Trial in Elderly Citizens. *PLoS ONE* **2015**, *10*, e0141641. [CrossRef]

39. Canto, J.G.; Rogers, W.J.; Goldberg, R.J.; Peterson, E.D.; Wenger, N.K.; Vaccarino, V.; Kiefe, C.I.; Frederick, P.D.; Sopko, G.; Zheng, Z.J.; et al. Association of age and sex with myocardial infarction symptom presentation and in-hospital mortality. *JAMA* **2012**, *307*, 813–822. [CrossRef]

 *nutrients*

Article

Selenoprotein-P Deficiency Predicts Cardiovascular Disease and Death

Lutz Schomburg [1], Marju Orho-Melander [2], Joachim Struck [3], Andreas Bergmann [3] and Olle Melander [2,4,*]

1   Institut für Experimentelle Endokrinologie, Charité—Universitätsmedizin Berlin, corporate member of Freie Universität Berlin, Humboldt-Universität zu Berlin, and Berlin Institute of Health, D-13353 Berlin, Germany
2   Department of Clinical Sciences, Malmö, Lund University, SE 214 28 Malmö, Sweden
3   Sphingotec GmbH, Neuendorfstrasse 15A, D-16761 Hennigsdorf, Germany
4   Department of Internal Medicine, Clinical Research Center, Skåne University Hospital, Jan Waldenströms gata 35, Bldg. 91, SE 214 28 Malmö, Sweden
*   Correspondence: olle.melander@med.lu.se; Tel.: +46-4039-1209

Received: 24 July 2019; Accepted: 7 August 2019; Published: 9 August 2019

Abstract: Selenoprotein-P (SELENOP) is the main carrier of selenium to target organs and reduces tissue oxidative stress both directly and by delivering selenium to protective selenoproteins. We tested if the plasma concentration of SELENOP predicts cardiovascular morbidity and mortality in the primary preventive setting. SELENOP was measured from the baseline exam in 2002–2006 of the Malmö Preventive Project, a population-based prospective cohort study, using a validated ELISA. Quintiles of SELENOP concentration were related to the risk of all-cause mortality, cardiovascular mortality, and a first cardiovascular event in 4366 subjects during a median (interquartile range) follow-up time of 9.3 (8.3–11) years using Cox proportional Hazards Model adjusting for cardiovascular risk factors. Compared to subjects in the lowest quintile of SELENOP, the risk of all three endpoints was significantly lower in quintiles 2–5. The risk (multivariate adjusted hazard ratio, 95% CI) decreased gradually with the lowest risk in quintile 4 for all-cause mortality (0.57, 0.48–0.69) ($p < 0.001$), cardiovascular mortality (0.52, 0.37–0.72) ($p < 0.001$), and first cardiovascular event (0.56, 0.44–0.71) ($p < 0.001$). The lower risk of a first cardiovascular event in quintiles 2–5 as compared to quintile 1 was significant for both coronary artery disease and stroke. We conclude that the 20% with lowest SELENOP concentrations in a North European population without history of cardiovascular disease have markedly increased risk of cardiovascular morbidity and mortality, and preventive selenium supplementation studies stratified for these subjects are warranted.

Keywords: Selenoprotein-P; selenium; cardiovascular disease; prevention; supplementation

---

1. Introduction

Selenium (Se) is an essential micronutrient of fundamental importance for human health [1]. A small group of proteins contains selenium as the 21st proteinogenic amino acid selenocysteine in their primary sequence at the active site, e.g., the glutathione peroxidases, thioredoxin reductases, iodothyronine deiodinases, and others [2]. These so-called selenoproteins are encoded by a set of 25 genes in humans [3]. Inherited defects causing reduced selenoprotein biosynthesis lead to a complex disease syndrome with myopathic features, male infertility, abnormal thyroid hormones, and signs of increased oxidative stress associated with high ultraviolet light (UV) radiation sensitivity and eventually neuronal loss [4].

The trace element is supplied by a regular diet; however, selenium intake levels differ strongly around the world due to biogeochemical differences [5]. Accordingly, the Se status of different populations depends on their soil quality as well as on the origin and pattern of the diet. Most of

Europe, Africa, and Asia are considered as insufficiently supplied in contrast to e.g., the United States or Canada, partly reflected in genetic adaptations in populations residing in very selenium poor areas [6].

A low selenium intake causes insufficient expression of selenoproteins, low Se concentrations in the circulation and tissues, as well as an increased risk for certain diseases including colorectal cancer [7], autoimmune thyroid disease [8] or a sub-responsive immune system [9]. Different biomarkers are available for Se status assessment, of which total serum or plasma selenium concentrations and selenoprotein P (SELENOP) levels are established and considered as most reliable [10,11]. Circulating SELENOP mainly derives from hepatocytes and serves a Se transport function. Target cells express SELENOP-receptors that belong to the family of lipoprotein receptor-related proteins and can thereby become preferentially supplied with the essential trace element under Se-deficient conditions [12]. In addition, SELENOP also shows enzymatic activity and may protect vascular endothelial cells from oxidative and nitrosative stress and damage [13].

However, SELENOP status has not been studied yet with respect to cardiovascular disease (CVD), and there is a lack of large prospective studies in the primary preventive setting for the potential relationship between SELENOP and mortality [14]. Given the role of SELENOP as functional marker of Se status and availability, and the need for a large enough group of healthy subjects with sub-optimal Se supply, we tested the association of SELENOP with cardiovascular disease in a large European prospective cohort study.

2. Materials and Methods

2.1. Study Population

The Malmö Preventive Project (MPP) is a Swedish single-center prospective population-based study. Between 1974–1992, in all 33,346 citizens of the city of Malmö in Southern Sweden were included. The recruited subjects were screened for traditional cardiovascular risk factors. Between 2002–2006, all subjects alive were invited for a re-examination in which 18,240 individuals participated. This re-examination in 2002–2006 forms the baseline exam of the current study. Here, cardiovascular risk factors were assessed, and plasma was separated and frozen to −80 degrees for later analyses. Approval was granted by the Regional Board of Ethics, Lund, Sweden (#2009/633). Of the 18,240 subjects, we excluded 2087 subjects who had had a cardiovascular disease event (coronary artery disease, myocardial infarction or stroke) prior to the baseline exam leaving 16,153 individuals. Of these, 15,743 had complete data on cardiovascular risk factors, from whom a random sample of 4500 subjects was selected for analysis of SELENOP, among whom 4366 had a stored EDTA-plasma available in which SELENOP was subsequently analyzed (Supplementary Figure S1).

All study subjects signed oral and written informed consent to participate and to publish the results, and the study protocols were approved by the Regional Board of Ethics in Lund, Sweden.

2.2. Clinical Examination and Assays

Participants underwent a medical history, physical examination, and laboratory assessment. Blood pressure was measured using an oscillometric device twice after 10 min of rest in the supine position. Diabetes mellitus was defined as fasting plasma glucose 7.0 mmol/L or above, a self-reported physician diagnosis of diabetes, or use of anti-diabetic medication. Cigarette smoking was elicited by a self-administered questionnaire, with current cigarette smoking defined as any use within the past year. Measurements of fasting serum total cholesterol, high density lipoprotein (HDL) cholesterol, and triglycerides were made according to standard procedures at the Department of Clinical Chemistry, Skåne University Hospital. low density lipoprotein (LDL) cholesterol was calculated according to Friedewald's formula. Plasma SELENOP concentrations were measured in fasted ethylene-diamine-tetraacetic acid (EDTA)-plasma using a validated ELISA (selenOtest ELISA, selenOmed GmbH, Berlin, Germany) characterized recently in detail [15], which was independently proven as a highly reliable commercial assay [16]. Corresponding selenium concentrations were measured in a subsample of 284 subjects using total reflection X-ray fluorescence (Picofox S2, Bruker nano) as described [7].

2.3. Endpoints

The baseline plasma concentration of SELENOP was analyzed in the study sample of 4366 individuals free of any of the primary outcomes in question in relation to three primary outcomes: a first cardiovascular event, all-cause mortality and cardiovascular mortality (for event definitions, see below). In secondary analyses, we separated the cardiovascular disease endpoint into incidence of coronary artery disease (CAD) and stroke. The endpoints were retrieved through record linkage of the personal identification number of each Swedish individual and the Swedish Hospital Discharge Register (SHDR), the Swedish Cause of Death Register (SCDR), the Stroke in Malmö Register, and the Swedish Coronary Angiography and Angioplasty Registry (SCAAR). These registers were previously described and validated for classification of outcomes [17–19]. CAD was defined as fatal or nonfatal myocardial infarction, death due to ischemic heart disease, percutaneous coronary intervention (PCI), or coronary artery bypass grafting (CABG), whichever came first. Stroke was defined as fatal or nonfatal stroke. Follow-up for outcomes extended to 31 December 2014. Cardiovascular death was defined a main cause of death diagnosis, according to the death certificate, between codes 390–459 of the International Classification of Disease (ICD) version 9 or within the I-chapter of ICD version 10.

2.4. Statistics

We measured the concentration of SELENOP in plasma from the baseline examination (between 2002–2006) of 4366 subjects of the Malmö Preventive Project who were free from prior cardiovascular disease. Quintiles of SELENOP (lowest quintile defined as reference) were related to risk of (1) all-cause mortality, (2) cardiovascular mortality, and (3) a first cardiovascular disease event (fatal or non-fatal myocardial infarction or stroke, coronary revascularization or death due to coronary heart disease) during follow-up using Cox Proportional Hazards Models adjusted for age, gender, current smoking, systolic blood pressure, use of antihypertensive medication, diabetes mellitus, LDL-cholesterol, HDL-cholesterol, and body mass index. Correlation between plasma concentration of SELENOP and Se was tested using Spearman's correlation. Statistical analysis was performed using SPSS (v22.0; IBM Corp., Armonk, NY, USA). A two-sided $p < 0.05$ was considered significant.

3. Results

Baseline characteristics of the study population, stratified for quintiles of baseline SELENOP plasma concentration, are shown in Table 1. The most evident difference according to baseline SELENOP quintile was a smoking prevalence of 28% in the lowest quintile of SELENOP as compared to 16–18% in the other four quintiles. Furthermore, there were slight but significant linear or non-linear differences between SELENOP quintiles for age, gender, diabetes mellitus, LDL-cholesterol, HDL-cholesterol, and body mass index (Table 1).

During a median (interquartile range) follow-up time of 9.3 (8.3–11) years, a total of 1111 deaths occurred. The largest number of deaths was observed in SELENOP quintile 1 ($n = 314$). The number of deaths decreased with higher quintiles and the lowest number of deaths was recorded in quintile 4 ($n = 175$), followed by a nominal increase in the number of deaths in quintile 5 ($n = 215$) (Table 2). In multivariate adjusted analyses, the risk of all-cause mortality was highly significantly lower in each of SELENOP quintiles 2–5 compared to the lowest SELENOP quintile with the lowest point estimate of the hazard ratio in quintile 4.

Table 1. Baseline clinical characteristics according to quintile (Q) of concentration of Selenoprotein-P (SELENOP) at baseline of the subjects analyzed who were without history of cardiovascular disease.

	Q1	Q2	Q3	Q4	Q5	*p*	*p*-Trend
	(*n* = 873)	(*n* = 873)	(*n* = 874)	(*n* = 873)	(*n* = 873)		
SELENOP [1] (mg/L)	3.7 (0.4–4.3)	4.7 (4.3–5.1)	5.5 (5.1–5.9)	6.3 (5.9–6.9)	7.7 (6.9–20)	n. a.	n. a.
Age (years)	70 ± 6.4	69 ± 6.4	69 ± 6.1	69 ± 6.1	70 ± 6.1	<0.001	0.01
Gender, *n* (%) male	594 (68)	641 (73)	587 (67)	633 (73)	553 (63)	<0.001	0.04
Current smoking, *n* (%)	240 (28)	157 (18)	149 (17)	146 (17)	143 (16)	<0.001	<0.001
Systolic blood pressure (mmHg)	147 ± 21	147 ± 20	146 ± 21	146 ± 20	147 ± 20	n. s.	n. s.
Antihypertensive medication, *n* (%)	311 (36)	277 (32)	321 (37)	289 (33)	278 (32)	n. s.	n. s.
Diabetes Mellitus, *n* (%)	86 (9.9)	83 (9.5)	79 (9.0)	103 (12)	115 (13)	0.024	0.007
LDL-cholesterol (mmol/L)	3.62 ± 0.98	3.73 ± 0.96	3.74 ± 0.93	3.74 ± 0.97	3.71 ± 0.99	0.043	n. s.
HDL-cholesterol (mmol/L)	1.37 ± 0.42	1.34 ± 0.39	1.36 ± 0.38	1.39 ± 0.39	1.44 ± 0.43	<0.001	<0.001
Body Mass Index (kg/m^2)	26.9 ± 4.6	27.3 ± 4.2	27.5 ± 4.3	27.0 ± 3.8	27.0 ± 4.1	0.034	n. s.

[1] SELENOP; plasma concentration of selenoprotein P, Q; quintile, LDL; low density lipoprotein, HDL; high density lipoprotein, n. a.; not applicable, n. s.; non-significant.

Table 2. Population quintile (Q) of SELENOP in relation to all-cause mortality, cardiovascular mortality and a first cardiovascular event in subjects without history of cardiovascular disease at baseline in multivariate adjusted models.

Parameter	Q1	Q2	Q3	Q4	Q5
	(*n* = 873)	(*n* = 873)	(*n* = 874)	(*n* = 873)	(*n* = 873)
SELENOP [1] (mg/L)	3.7 (0.4–4.3)	4.7 (4.3–5.1)	5.5 (5.1–5.9)	6.3 (5.9–6.9)	7.7 (6.9–20)
ALL-CAUSE MORTALITY					
Number of events	314	214	193	175	215
Hazard Ratio (95% CI)	1.0 (ref)	0.73 *** (0.61–0.87)	0.66 *** (0.55–0.79)	0.57 *** (0.48–0.69)	0.69 *** (0.58–0.82)
CARDIOVASCULAR MORTALITY					
Number of events	106	66	66	53	60
Hazard Ratio (95% CI)	1.0 (ref)	0.65 ** (0.48–0.89)	0.66 ** (0.48–0.89)	0.52 *** (0.37–0.72)	0.59 ** (0.43–0.81)
FIRST CARDIOVASCULAR EVENT					
Number of events	188	157	145	115	140
Hazard Ratio (95% CI)	1.0 (ref)	0.79 * (0.64–0.98)	0.75 * (0.61–0.94)	0.56 *** (0.44–0.71)	0.70 ** (0.56–0.87)

[1] SELENOP; plasma concentration of selenoprotein P; CI, confidence interval. All analyses were adjusted for age, gender, current smoking, systolic blood pressure, use of antihypertensive medication, diabetes mellitus, LDL-cholesterol, HDL-cholesterol, and body mass index. * $p < 0.05$; ** $p < 0.01$; *** $p < 0.001$.

Similar patterns were observed for the crude and multivariate adjusted endpoint analyses of cardiovascular mortality (345 events) and risk of a first cardiovascular event (745 events), respectively, with significantly lower risks in each of SELENOP quintiles 2–5 compared to the bottom SELENOP quintile and the lowest point estimate of risk in SELENOP quintile 4 (Table 2). The individuals of each of SELENOP quintiles 2–5 had significantly lower risks as compared to the individuals of the lowest SELENOP quintile Q1 (Figure 1).

For this reason, we subsequently compared subjects of the lowest SELENOP quintile (SELENOP deficiency) with all subjects of SELENOP quintile 2–5 (normal SELENOP). In multivariate adjusted analyses, subjects with SELENOP deficiency as compared to subjects with normal SELENOP plasma concentration had a hazard ratio (95% confidence interval) of 1.51 (1.32–1.72) ($p = 1.2 \times 10^{-9}$) for all-cause mortality; 1.61 (1.32–2.09) ($p = 1.7 \times 10^{-5}$) for cardiovascular mortality, and 1.43 (1.21–1.69) ($p = 2.7 \times 10^{-5}$) for first cardiovascular event. When breaking up cardiovascular events into its two components, subjects with SELENOP deficiency were at significantly increased risk of both coronary artery disease (490 events) [1.27 (1.03–1.57) ($p = 0.025$)] and stroke (305 events) [1.57 (1.21–2.02) ($p = 0.001$)].

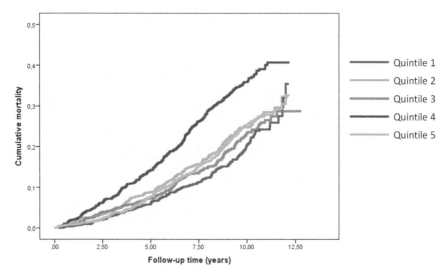

Figure 1. Kaplan Meier analysis for mortality risk in relation to Selenoprotein-P (SELENOP) status. Over the course of up to 12.5 years, the cumulative rates of mortality differed between the lowest quintile (Q1) of SELENOP plasma concentrations and the higher quintiles (Q2–Q5). A quantitative analysis is found in Table 2.

In the subsample in which both SELENOP and selenium was measured there was significant correlation between the two parameters of Se status (R = 0.66, $p = 4.0 \times 10^{-37}$) (Figure 2).

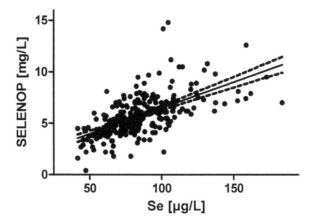

Figure 2. Correlation analysis between SELENOP and Se concentrations. A subset of 284 plasma samples was analyzed for both SELENOP and Se concentrations. The two biomarkers of Se status correlate strongly across the study cohort, indicative of sub-optimal Se intake (Spearman's correlation coefficient; r = 0.6604).

Even though all analyses were adjusted for cardiovascular risk factors including smoking, we subsequently performed stratified analyses in smokers and non-smokers in order to make sure the elevated risk for cardiovascular morbidity and mortality in subjects with SELENOP deficiency was not caused by their higher smoking rates (Table 1).

Interestingly, among non-smokers SELENOP deficiency was significantly associated with all-cause mortality [1.56 (1.33–1.82) ($p = 2.9 \times 10^{-8}$)], cardiovascular mortality [1.88 (1.44–2.45) ($p = 3.0 \times 10^{-6}$)] and first cardiovascular event [1.47 (1.21–1.79) ($p = 1.1 \times 10^{-4}$)], whereas the association between SELENOP deficiency and the three main endpoints was weaker or non-significant among smokers (Table 3).

Table 3. Association of SELENOP status in relation to major endpoints in non-smokers vs. smokers.

Parameter	Non-Smokers ($n = 3531$)		Smokers ($n = 835$)	
SELENOP-Deficient vs. Normal	**Hazard Ratio (95% CI)**	***p*-Value**	**Hazard Ratio (95% CI)**	***p*-Value**
All-cause mortality	1.56 (1.33–1.82)	<0.001	1.35 (1.05–1.74)	0.018
CVD mortality	1.88 (1.44–2.45)	<0.001	1.23 (0.70–1.80)	NS
First CVD event	1.47 (1.21–1.79)	<0.001	1.32 (0.95–1.82)	NS

CVD, cardiovascular disease.

4. Discussion

We report a strong association between low SELENOP concentrations and the risk for all-cause mortality, cardiovascular mortality and a first cardiovascular event in a large group of adult Swedish subjects with no history of cardiovascular events prior to baseline, i.e., in a primary preventive setting. The low at-risk quintile (SELENOP Q1) identified is characterized by serum SELENOP concentrations below 4.3 mg/L SELENOP, corresponding to serum Se concentrations of less than 70 µg/L. These thresholds for the lowest quintile are similar to the corresponding values determined in the European prospective investigation of cancer and nutrition cohort (EPIC) [7]. Here, an analysis of 966 patients and 966 matched controls from eight different European countries identified an increased risk for colorectal cancer in the lowest quintiles of SELENOP and selenium concentrations, respectively, i.e., below a concentration of 3.6 mg/L of SELENOP or below 67.7 µg/L of total Se [7]. The slightly lower boundaries of SELENOP Q1 in the EPIC analysis may be related to the tendency that Northern European populations are better supplied with selenium than the subjects in central or southern parts of Europe [7].

A total serum selenium concentration in the range of 70 µg/L is known to indicate a sub-optimal expression of circulating selenoproteins including the glutathione peroxidases and SELENOP [10,20,21]. Full expression of SELENOP requires higher selenium intakes than that required for GPX1 or GPX3 saturation, and SELENOP is therefore considered as the most suitable protein-based biomarker of Se status becoming maximally expressed at serum or plasma selenium concentrations of 125 µg/L [10,20,21]. This concentration is found only in very few subjects of the population studied, and a linear association of plasma selenium concentrations with SELENOP levels is observed, indicating deficiency (Figure 2). In general, serum selenium concentrations of 125 µg/L or more are rarely observed in Europe, and a considerable fraction of the population is considered as selenium- deficient. The intake required for reaching a selenium status that might provide optimal protection from selenium-deficiency related diseases is unknown, but a U-shaped interaction between health risks or benefits and Se status is widely accepted [1,22,23]. Interestingly, Finland started a population-wide selenium supplementation effort more than 30 years ago and raised the average plasma selenium concentration from around 70 µg/L in 1985 to current levels of around 111 µg/L [24]. Our data suggest that this decision was most likely taken wisely, as hereby many subjects will have been promoted from the at risk quintile Q1 determined in this study to a higher SELENOP status. Yet, our study is reporting associations only, and should not be mis-interpreted as proof of causality.

Our results contribute a novel aspect to the abundant literature on selenium status and cardiovascular disease [25]. Specifically, we provide evidence on the potential relevance of the selenium transporter SELENOP in relation to cardiovascular morbidity and mortality. SELENOP may modify cardiovascular disease risk by several mechanisms [26]: It transports selenium to vital tissues that are equipped with receptors (megalin or APOER2) for SELENOP uptake, thereby increasing intracellular selenoprotein biosynthesis for improving antioxidative defense and protein quality control systems [12]. SELENOP exhibits GPX activity and is capable of catalyzing degradation of phospholipid hydroperoxides, thereby protecting cell membrane integrity [27] and LDL-particles from oxidation [28]. SELENOP is also known to reducing peroxynitrite [29], and to associate with the extracellular matrix via a heparin binding domain [30]. Finally SELENOP binds heavy metals like Cd, As, and Hg thereby reducing oxidative stress and avoiding toxic damage in the circulation [31]. Especially the latter notion has been supported by a recent study with Hg-exposed Inuit, where subjects with high selenium intake and status were less hypertensive and displayed reduced stroke and myocardial infarction rates as compared to those with a lower selenium status [32].

Our results align with prior studies on the inverse relation of certain selenoproteins with cardiovascular disease risk. In a prospective study of >600 patients with suspected coronary artery disease, GPX1 erythrocyte activity was related to the risk of cardiovascular events, independent from smoking status [33]. Similarly, circulating levels of the extracellular GPX isoform (GPX3) were inversely related to the risk of cardiovascular events in patients with atrial fibrillation in a prospective cohort study with 909 patients [34]. Notably, both studies had been performed in countries with insufficient selenium intake, i.e., Germany and Italy, respectively. The recent data from the Minnesota Heart Survey also indicate that selenium status in the form of GPX3 activity is inversely correlated to cardiovascular disease mortality even in a selenium replete population [35]. GPX3 is a valid biomarker for chronic kidney disease, contributing to overall selenium status and affecting systemic oxidative stress [36]. Notably, renal GPX3 expression depends on liver-derived SELENOP [37], and SELENOP should thus be considered as a more direct and reliable biomarker of selenium status [38].

While an inverse relation between selenium status and cardiovascular disease risk is found in most of the clinical studies, the results from intervention trials are ambiguous [39]. A recent meta-analysis showed that selenium supplementation does not generally reduce cardiovascular disease risk, probably due to the inclusion of results from studies conducted in areas with relatively high baseline selenium status without selecting individuals with low selenium status [40].

In combination with the findings from selenium-replete subjects where a positive interaction of very high selenium status with hypertension has been observed [41], and lowest mortality risk is seen in subjects with lowest Se levels [42], our data reinforce the idea of a U-shaped interaction between selenium status and mortality risk. Specifically, our study highlights the lower boundary of selenium intake and selenium status (Figure 3). The cardiovascular disease and mortality risks of the majority of our study subjects were independent of the selenium status, indicating that their selenium status was within the plateau phase connecting selenium-deficiency from selenium-oversupply. However, about 20% of subjects, i.e., the ones residing in the lowest quintile Q1, exhibited a strongly increased health risk and may profit from supplemental selenium.

This notion has several potential clinical implications: (1) Subjects with potentially low selenium intake should be tested for SELENOP deficiency and advised with respect to taking natural selenium rich products or supplements. (2) There is a need for randomized controlled trials (RCT) in selenium-deficient populations specifically in subjects with SELENOP concentrations corresponding to SELENOP Q1, to verify that selenium-containing supplements or a selenium-rich diet can increase SELENOP levels and thereby reduce cardiovascular disease risk in these subjects. (3) Natural, environmental and pharmacological modifiers of SELENOP expression need to be identified in order to better control selenium status and be tested in relation to cardiovascular disease risk. Our study has limitations. Due to the observational nature of the study, we cannot prove that the associations between SELENOP and the study endpoints are causal. For this, RCTs targeting the low SELENOP segment of the population are needed. Moreover, the MPP included more men than women and our study population, surveyed 2002–2006, represents survivors from the original baseline examination 1974–1992 and thus the subjects enrolled are likely healthier than the background population.

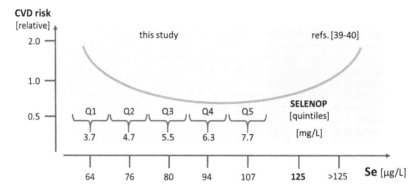

Figure 3. Presumed U-shaped interaction of Se status and CVD risk. The present study indicates a highly increased risk for cardiovascular endpoints in subjects residing in the lowest quintile (Q1) of SELENOP status as compared to the other subjects with higher SELENOP status (Q2–Q5). The figure presents both the plasma SELENOP and corresponding total Se concentrations. The green shaded area denotes the results from the current study, whereas the red shaded part presents an extrapolation of published studies from areas with higher baseline selenium status.

5. Conclusions

We conclude that in a North European population without history of cardiovascular disease, the 20% with lowest SELENOP concentrations have markedly increased risk of cardiovascular morbidity and mortality. Rather than population-wide supplementation strategies, clinical trials testing if cardiovascular morbidity and mortality can be reduced in subjects belonging to this low SELENOP stratum, are warranted.

6. Patents

Dr. Bergmann reports being president of Sphingotec GmbH, which holds the patent rights for the use of Selenoprotein-P in prediction of cardiovascular disease. Prof. Melander reports being listed as inventor on the same patent application.

Supplementary Materials: The following are available online at http://www.mdpi.com/2072-6643/11/8/1852/s1, Figure S1: Consort Diagram.

Author Contributions: O.M., M.O.-M., and L.S. designed the study. O.M. and M.O.-M. did the statistical analyses. O.M. and L.S. wrote the manuscript. O.M., L.S., J.S., and A.B. collected the material. All authors critically reviewed the manuscript and approved its submission.

Funding: Melander was supported by grants from Knut and Alice Wallenberg Foundation, Göran Gustafsson Foundation, the Swedish Heart- and Lung Foundation, the Swedish Research Council, the Novo Nordisk Foundation, Region Skåne, Skåne University Hospital and the Swedish Foundation for Strategic Research Dnr IRC15-0067. Schomburg was supported by grants from Charité Medical School Berlin and the Deutsche Forschungsgemeinschaft (DFG Research Unit 2558 TraceAge, Scho 849/6-1). Orho-Melander was funded by the European Research Council (consolidator grant 649021), the Swedish Research Council, the Novo Nordisk Foundation, the European Foundation for the Study of Diabetes (EFSD 2015-338), Region Skåne, Skåne University Hospital and the Swedish Foundation for Strategic Research Dnr IRC15-0067 and the Swedish Heart and Lung Foundation.

Conflicts of Interest: Struck reports being employed by Sphingotec GmbH. Schomburg holds shares in selenOmed GmbH, a company involved in selenium status assessment and supplementation. The funders stated above had no role in the design of the study; in the collection, analyses, or interpretation of data; in the writing of the manuscript, or in the decision to publish the results.

References

1. Rayman, M.P. Selenium and human health. *Lancet (Lond. Engl.)* **2012**, *379*, 1256–1268. [CrossRef]
2. Hatfield, D.L.; Tsuji, P.A.; Carlson, B.A.; Gladyshev, V.N. Selenium and selenocysteine: Roles in cancer, health, and development. *Trends Biochem. Sci.* **2014**, *39*, 112–120. [CrossRef] [PubMed]
3. Kryukov, G.V.; Castellano, S.; Novoselov, S.V.; Lobanov, A.V.; Zehtab, O.; Guigo, R.; Gladyshev, V.N. Characterization of mammalian selenoproteomes. *Science* **2003**, *300*, 1439–1443. [CrossRef] [PubMed]
4. Schoenmakers, E.; Agostini, M.; Mitchell, C.; Schoenmakers, N.; Papp, L.; Rajanayagam, O.; Padidela, R.; Ceron-Gutierrez, L.; Doffinger, R.; Prevosto, C.; et al. Mutations in the selenocysteine insertion sequence-binding protein 2 gene lead to a multisystem selenoprotein deficiency disorder in humans. *J. Clin. Investig.* **2010**, *120*, 4220–4235. [CrossRef] [PubMed]
5. Winkel, L.H.; Vriens, B.; Jones, G.D.; Schneider, L.S.; Pilon-Smits, E.; Banuelos, G.S. Selenium cycling across soil-plant-atmosphere interfaces: A critical review. *Nutrients* **2015**, *7*, 4199–4239. [CrossRef] [PubMed]
6. White, L.; Romagne, F.; Muller, E.; Erlebach, E.; Weihmann, A.; Parra, G.; Andres, A.M.; Castellano, S. Genetic adaptation to levels of dietary selenium in recent human history. *Mol. Biol. Evol.* **2015**, *32*, 1507–1518. [CrossRef]
7. Hughes, D.J.; Fedirko, V.; Jenab, M.; Schomburg, L.; Meplan, C.; Freisling, H.; Bueno-de-Mesquita, H.B.; Hybsier, S.; Becker, N.P.; Czuban, M.; et al. Selenium status is associated with colorectal cancer risk in the European prospective investigation of cancer and nutrition cohort. *Int. J. Cancer* **2015**, *136*, 1149–1161. [CrossRef]
8. Wu, Q.; Rayman, M.P.; Lv, H.; Schomburg, L.; Cui, B.; Gao, C.; Chen, P.; Zhuang, G.; Zhang, Z.; Peng, X.; et al. Low Population Selenium Status Is Associated with Increased Prevalence of Thyroid Disease. *J. Clin. Endocrinol. Metab.* **2015**, *100*, 4037–4047. [CrossRef]
9. Avery, J.C.; Hoffmann, P.R. Selenium, Selenoproteins, and Immunity. *Nutrients* **2018**, *10*, 1203. [CrossRef]
10. Hurst, R.; Armah, C.N.; Dainty, J.R.; Hart, D.J.; Teucher, B.; Goldson, A.J.; Broadley, M.R.; Motley, A.K.; Fairweather-Tait, S.J. Establishing optimal selenium status: Results of a randomized, double-blind, placebo-controlled trial. *Am. J. Clin. Nutr.* **2010**, *91*, 923–931. [CrossRef]
11. Combs, G.F., Jr.; Watts, J.C.; Jackson, M.I.; Johnson, L.K.; Zeng, H.; Scheett, A.J.; Uthus, E.O.; Schomburg, L.; Hoeg, A.; Hoefig, C.S.; et al. Determinants of selenium status in healthy adults. *Nutr. J.* **2011**, *10*, 75. [CrossRef] [PubMed]
12. Burk, R.F.; Hill, K.E. Regulation of Selenium Metabolism and Transport. *Annu. Rev. Nutr.* **2015**, *35*, 109–134. [CrossRef] [PubMed]
13. Steinbrenner, H.; Sies, H. Protection against reactive oxygen species by selenoproteins. *Biochim. Biophys. Acta* **2009**, *1790*, 1478–1485. [CrossRef] [PubMed]
14. Benstoem, C.; Goetzenich, A.; Kraemer, S.; Borosch, S.; Manzanares, W.; Hardy, G.; Stoppe, C. Selenium and Its Supplementation in Cardiovascular Disease-What do We Know? *Nutrients* **2015**, *7*, 3094–3118. [CrossRef] [PubMed]
15. Hybsier, S.; Schulz, T.; Wu, Z.; Demuth, I.; Minich, W.B.; Renko, K.; Rijntjes, E.; Kohrle, J.; Strasburger, C.J.; Steinhagen-Thiessen, E.; et al. Sex-specific and inter-individual differences in biomarkers of selenium status identified by a calibrated ELISA for selenoprotein P. *Redox Biol.* **2017**, *11*, 403–414. [CrossRef] [PubMed]
16. Saito, Y.; Misu, H.; Takayama, H.; Takashima, S.; Usui, S.; Takamura, M.; Kaneko, S.; Takamura, T.; Noguchi, N. Comparison of Human Selenoprotein P Determinants in Serum between Our Original Methods and Commercially Available Kits. *Biol. Pharm. Bull.* **2018**, *41*, 828–833. [CrossRef]
17. Hammar, N.; Alfredsson, L.; Rosen, M.; Spetz, C.L.; Kahan, T.; Ysberg, A.S. A national record linkage to study acute myocardial infarction incidence and case fatality in Sweden. *Int. J. Epidemiol.* **2001**, *30* (Suppl. 1), S30–S34. [CrossRef]
18. Jerntorp, P.; Berglund, G. Stroke registry in Malmo, Sweden. *Stroke* **1992**, *23*, 357–361. [CrossRef]
19. Sarno, G.; Lagerqvist, B.; Carlsson, J.; Olivecrona, G.; Nilsson, J.; Calais, F.; Gotberg, M.; Nilsson, T.; Sjogren, I.; James, S. Initial clinical experience with an everolimus eluting platinum chromium stent (Promus Element) in unselected patients from the Swedish Coronary Angiography and Angioplasty Registry (SCAAR). *Int. J. Cardiol.* **2013**, *167*, 146–150. [CrossRef]

20. Xia, Y.; Hill, K.E.; Li, P.; Xu, J.; Zhou, D.; Motley, A.K.; Wang, L.; Byrne, D.W.; Burk, R.F. Optimization of selenoprotein P and other plasma selenium biomarkers for the assessment of the selenium nutritional requirement: A placebo-controlled, double-blind study of selenomethionine supplementation in selenium-deficient Chinese subjects. *Am. J. Clin. Nutr.* **2010**, *92*, 525–531. [CrossRef]

21. Xia, Y.M.; Hill, K.E.; Byrne, D.W.; Xu, J.Y.; Burk, R.F. Effectiveness of selenium supplements in a low-selenium area of China. *Am. J. Clin. Nutr.* **2005**, *81*, 829–834. [CrossRef] [PubMed]

22. Chiang, E.C.; Shen, S.R.; Kengeri, S.S.; Xu, H.P.; Combs, G.F.; Morris, J.S.; Bostwick, D.G.; Waters, D.J. Defining the Optimal Selenium Dose for Prostate Cancer Risk Reduction: Insights from the U-Shaped Relationship between Selenium Status, DNA Damage, and Apoptosis. *Dose-Response* **2010**, *8*, 285–300. [CrossRef] [PubMed]

23. Bleys, J.; Navas-Acien, A.; Guallar, E. Serum selenium levels and all-cause, cancer, and cardiovascular mortality among US adults. *Arch. Intern. Med.* **2008**, *168*, 404–410. [CrossRef] [PubMed]

24. Alfthan, G.; Eurola, M.; Ekholm, P.; Venalainen, E.R.; Root, T.; Korkalainen, K.; Hartikainen, H.; Salminen, P.; Hietaniemi, V.; Aspila, P.; et al. Effects of nationwide addition of selenium to fertilizers on foods, and animal and human health in Finland: From deficiency to optimal selenium status of the population. *J. Trace Elem. Med. Biol.* **2015**, *31*, 142–147. [CrossRef] [PubMed]

25. Stranges, S.; Navas-Acien, A.; Rayman, M.P.; Guallar, E. Selenium status and cardiometabolic health: State of the evidence. *Nutr. Metab. Cardiovasc. Dis.* **2010**, *20*, 754–760. [CrossRef] [PubMed]

26. Saito, Y.; Sato, N.; Hirashima, M.; Takebe, G.; Nagasawa, S.; Takahashi, K. Domain structure of bi-functional selenoprotein P. *Biochem. J.* **2004**, *381*, 841–846. [CrossRef] [PubMed]

27. Saito, Y.; Hayashi, T.; Tanaka, A.; Watanabe, Y.; Suzuki, M.; Saito, E.; Takahashi, K. Selenoprotein P in human plasma as an extracellular phospholipid hydroperoxide glutathione peroxidase—Isolation and enzymatic characterization of human selenoprotein P. *J. Biol. Chem.* **1999**, *274*, 2866–2871. [CrossRef]

28. Traulsen, H.; Steinbrenner, H.; Buchczyk, D.P.; Klotz, L.O.; Sies, H. Selenoprotein P protects low-density lipoprotein against oxidation. *Free Radic. Res.* **2004**, *38*, 123–128. [CrossRef]

29. Arteel, G.E.; Mostert, V.; Oubrahim, H.; Briviba, K.; Abel, J.; Sies, H. Protection by selenoprotein P in human plasma against peroxynitrite-mediated oxidation and nitration. *Biol. Chem.* **1998**, *379*, 1201–1205.

30. Hondal, R.J.; Ma, S.G.; Caprioli, R.M.; Hill, K.E.; Burk, R.F. Heparin-binding histidine and lysine residues of rat selenoprotein P. *J. Biol. Chem.* **2001**, *276*, 15823–15831. [CrossRef]

31. Sasakura, C.; Suzuki, K.T. Biological interaction between transition metals (Ag, Cd and Hg), selenide/sulfide and selenoprotein P. *J. Inorg. Biochem.* **1998**, *71*, 159–162. [CrossRef]

32. Hu, X.F.; Eccles, K.M.; Chan, H.M. High selenium exposure lowers the odds ratios for hypertension, stroke, and myocardial infarction associated with mercury exposure among Inuit in Canada. *Environ. Int.* **2017**, *102*, 200–206. [CrossRef] [PubMed]

33. Blankenberg, S.; Rupprecht, H.J.; Bickel, C.; Torzewski, M.; Hafner, G.; Tiret, L.; Smieja, M.; Cambien, F.; Meyer, J.; Lackner, K.J.; et al. Glutathione peroxidase 1 activity and cardiovascular events in patients with coronary artery disease. *N. Engl. J. Med.* **2003**, *349*, 1605–1613. [CrossRef] [PubMed]

34. Pastori, D.; Pignatelli, P.; Farcomeni, A.; Menichelli, D.; Nocella, C.; Carnevale, R.; Violi, F. Aging-Related Decline of Glutathione Peroxidase 3 and Risk of Cardiovascular Events in Patients with Atrial Fibrillation. *J. Am. Heart Assoc.* **2016**, *5*, e003682. [CrossRef] [PubMed]

35. Buijsse, B.; Lee, D.H.; Steffen, L.; Erickson, R.R.; Luepker, R.V.; Jacobs, D.R.; Holtzman, J.L. Low Serum Glutathione Peroxidase Activity Is Associated with Increased Cardiovascular Mortality in Individuals with Low HDLc's. *PLoS ONE* **2012**, *7*, e38901. [CrossRef] [PubMed]

36. Pang, P.; Abbott, M.; Abdi, M.; Fucci, Q.A.; Chauhan, N.; Mistri, M.; Proctor, B.; Chin, M.; Wang, B.; Yin, W.Q.; et al. Pre-clinical model of severe glutathione peroxidase-3 deficiency and chronic kidney disease results in coronary artery thrombosis and depressed left ventricular function. *Nephrol. Dial. Transpl.* **2018**, *33*, 923–934. [CrossRef] [PubMed]

37. Renko, K.; Werner, M.; Renner-Muller, I.; Cooper, T.G.; Yeung, C.H.; Hollenbach, B.; Scharpf, M.; Kohrle, J.; Schomburg, L.; Schweizer, U. Hepatic selenoprotein P (SePP) expression restores selenium transport and prevents infertility and motor-incoordination in Sepp-knockout mice. *Biochem. J.* **2008**, *409*, 741–749. [CrossRef]

38. Hargreaves, M.K.; Liu, J.G.; Buchowski, M.S.; Patel, K.A.; Larson, C.O.; Schlundt, D.G.; Kenerson, D.M.; Hill, K.E.; Burk, R.F.; Blot, W.J. Plasma Selenium Biomarkers in Low Income Black and White Americans from the Southeastern United States. *PLoS ONE* **2014**, *9*, e84972. [CrossRef]

39. Liu, H.M.; Xu, H.B.; Huang, K.X. Selenium in the prevention of atherosclerosis and its underlying mechanisms. *Metallomics* **2017**, *9*, 21–37. [CrossRef]

40. Zhang, X.; Liu, C.; Guo, J.; Song, Y. Selenium status and cardiovascular diseases: Meta-analysis of prospective observational studies and randomized controlled trials. *Eur. J. Clin. Nutr.* **2016**, *70*, 162–169. [CrossRef]

41. Laclaustra, M.; Navas-Acien, A.; Stranges, S.; Ordovas, J.M.; Guallar, E. Serum Selenium Concentrations and Hypertension in the US Population. *Circ. Cardiovasc. Qual.* **2009**, *2*, 369–376. [CrossRef] [PubMed]

42. Christensen, K.; Werner, M.; Malecki, K. Serum selenium and lipid levels: Associations observed in the National Health and Nutrition Examination Survey (NHANES) 2011–2012. *Environ. Res.* **2015**, *140*, 76–84. [CrossRef] [PubMed]

 *nutrients*

Article

Agrp-Specific Ablation of Scly Protects against Diet-Induced Obesity and Leptin Resistance

Daniel J. Torres, Matthew W. Pitts, Ann C. Hashimoto and Marla J. Berry *

Department of Cell and Molecular Biology, John A. Burns School of Medicine, University of Hawai'i, Honolulu, HI 96813, USA
* Correspondence: mberry@hawaii.edu; Tel.: +1-808-692-1506

Received: 28 June 2019; Accepted: 20 July 2019; Published: 23 July 2019

Abstract: Selenium, an essential trace element known mainly for its antioxidant properties, is critical for proper brain function and regulation of energy metabolism. Whole-body knockout of the selenium recycling enzyme, selenocysteine lyase (Scly), increases susceptibility to metabolic syndrome and diet-induced obesity in mice. Scly knockout mice also have decreased selenoprotein expression levels in the hypothalamus, a key regulator of energy homeostasis. This study investigated the role of selenium in whole-body metabolism regulation using a mouse model with hypothalamic knockout of Scly. Agouti-related peptide (Agrp) promoter-driven Scly knockout resulted in reduced weight gain and adiposity while on a high-fat diet (HFD). Scly-Agrp knockout mice had reduced Agrp expression in the hypothalamus, as measured by Western blot and immunohistochemistry (IHC). IHC also revealed that while control mice developed HFD-induced leptin resistance in the arcuate nucleus, Scly-Agrp knockout mice maintained leptin sensitivity. Brown adipose tissue from Scly-Agrp knockout mice had reduced lipid deposition and increased expression of the thermogenic marker uncoupled protein-1. This study sheds light on the important role of selenium utilization in energy homeostasis, provides new information on the interplay between the central nervous system and whole-body metabolism, and may help identify key targets of interest for therapeutic treatment of metabolic disorders.

Keywords: agrp; hypothalamus; leptin; scly; selenium; selenoprotein; thermogenesis; type 2 diabetes

1. Introduction

The trace element selenium (Se) is a vital micronutrient that promotes redox balance and protects cells from oxidative stress. Selenium is required for the synthesis of selenoproteins, which function in a wide range of biological processes, such as thyroid hormone metabolism [1], fertility [2], and inflammation [3]. While altered Se status and selenoprotein expression have been associated with metabolic disorders such as type 2 diabetes (T2D) and obesity in humans, the mechanisms underlying this relationship are not well understood [4]. In animal models, various metabolic disturbances can be induced by the targeted disruption of different selenoproteins [5–8], as well as the enzyme selenocysteine lyase (Scly) [9], which is important for intracellular Se utilization [10]. Scly catalyzes the breakdown of the amino acid selenocysteine into alanine and selenide, to be used for de novo selenoprotein synthesis [11,12]. Genetic deletion of Scly in mice results in a striking metabolic phenotype that includes glucose insensitivity and hyperinsulinemia, as well as a greater propensity to develop metabolic syndrome, with males exhibiting more drastic changes [9]. Scly knockout (KO) mice are also more susceptible to high-fat diet (HFD)-induced obesity and related complications [13]. Selenoprotein expression is reduced in several tissues of Scly KO mice, including the hypothalamus, a brain region involved in the regulation of energy metabolism [14]. Growing evidence depicts an association between hypothalamic damage and obesity in both humans and rodents [15].

The importance of Se for proper brain function is well documented [16] and a prominent role for selenoproteins in the hypothalamus is supported by several recent studies. Tagashita et al. demonstrated that broad hypothalamic KO of the selenocysteine-tRNA (Trsp) in mice causes an overweight phenotype [17]. Schriever et al. found that conditional KO of the selenoprotein glutathione peroxidase 4 (GPx4) in agouti-related peptide (Agrp)-positive neurons in the hypothalamus exacerbates diet-induced obesity [18]. Additionally, selenoprotein M (SelM) was found to support hypothalamic leptin signaling [19], which may underlie the obesity phenotype observed in SelM KO mice [7]. Circulating leptin acts upon neurons in the hypothalamus, including inhibiting Agrp neurons, to promote a negative energy balance. Resistance to the anorexigenic actions of leptin is one of the hallmarks of obesity and is strongly associated with T2D [20]. Hypothalamic oxidative stress and endoplasmic reticulum (ER) stress have been implicated as causative factors in leptin resistance development [21], implying that Se may play an important role in maintaining leptin sensitivity. Hypothalamic dysfunction, including impaired leptin signaling, may therefore contribute to the metabolic phenotype of Scly KO mice.

Known as 'first-order neurons' in sensing energy homeostasis signals to the brain, Agrp neurons are located within the arcuate nucleus (Arc) and the median eminence (ME) of the hypothalamus, placing them in close proximity to circulating hormones and nutrients in the bloodstream [22]. Interestingly, Agrp neurons have been shown to be particularly susceptible to developing leptin resistance compared to other neuron types [23]. To explore the potential contributions of hypothalamic dysfunction to the metabolic phenotype of whole-body Scly KO mice, we generated a mouse line with Cre-driven Agrp neuron-specific KO of Scly (Scly-Agrp KO mice). Male and female mice were investigated in parallel to characterize any potential sex difference. The results discussed in this report highlight an important role for hypothalamic Scly in mediating HFD-induced weight gain and leptin resistance in mice.

2. Materials and Methods

2.1. Animals

Agrp$^{tm1(Cre)Lowl}$/J mice with IRES inserted in exon 3 of the Agrp gene were purchased from The Jackson Laboratory (Bar Harbor, ME, USA) [24]. Agrp$^{tm1(cre)Lowl}$/J mice were cross-bred with C57/BL6J mice with loxP sites flanking the Scly gene (Scly$^{fl/fl}$) to generate male and female Scly$^{fl/fl; Agrp-Cre}$ mice (Scly-Agrp KO mice). Littermates with the Scly gene floxed, but lacking the cre-driver (Scly$^{fl/fl; Agrp-WT}$) were used as controls. Comparison of control male mice (Scly$^{fl/fl; Agrp-WT}$) versus mice containing the cre-driver without the Scly gene floxed (Scly$^{WT; Agrp-Cre}$) demonstrated that the cre-driver did not impact HFD-induced weight gain. Although female Scly$^{WT; Agrp-Cre}$ mice gained weight at a faster rate than Scly$^{fl/fl; Agrp-WT}$ mice, this trend was in opposition to the effect observed in experimental (Scly$^{fl/fl; Agrp-Cre}$; Scly-Agrp KO) mice and, thus, did not account for the main findings of the study.

Mice were maintained on a 12-h light/dark cycle and allowed ad libitum food and water access. Sterilized glucose, leptin, and vehicle control (phosphate-buffered saline) were administered via intraperitoneal (i.p.) bolus injection. Mice were either anaesthetized with tribromoethanol via i.p. injection prior to transcardial perfusion or euthanized via CO_2 asphyxiation prior to fresh tissue collection. All animal experiments and procedures were conducted with the approval of the University of Hawaii's Institutional Animal Care and Use Committee (IACUC). Animal Care and Use Committee (IACUC) Protocol: APN 09-871-9, approved: 16 August 2018. Institutional Biosafety Committee (IBC) Protocol: IBC #18-10-544-02-4A-1R, approved: 23 October 2018.

2.2. Experimental Design

Mice were weaned at 21 days, fed standard lab chow until 4 weeks of age, then switched to a high-fat diet (HFD) chow containing 45% kcal fat and an energy density of 4.7 kcal/g (Research Diets, New Brunswick, NJ, USA; D12451). Body weight was measured at 10:00 every 14 days thereafter. At 16 weeks of age, in vivo metabolic phenotyping was carried out using metabolic chambers to monitor food intake and respiratory metabolism. A glucose tolerance test (GTT) was performed at 18 weeks of age and mice were sacrificed at 24 weeks of age for tissue collection. Mice were fasted overnight for 16 h (18:00 until 10:00 the next day) and injected intraperitoneally with either leptin (1 mg/kg body weight; R & D Systems, Minneapolis, MN, USA; 498-OB) or the vehicle (sterilized phosphate-buffered saline) without leptin 1 h prior to sacrifice. Tissue was either fixed via perfusion for immunohistochemical assays or snap-frozen in liquid nitrogen for Western blot analysis.

2.3. Metabolic Chambers

Food intake, activity, and respiratory metabolism were measured in mice using the PanLab Oxylet*Pro*^TM System (Harvard Apparatus, Barcelona, Spain) per the manufacturer's instructions. Mice were placed in individual homecage-like chambers, with fresh bedding, food, and water, and allowed to acclimate for 24 h, followed by 48 h of data collection. Cage air was sampled for 7-mi epochs every 35 min to measure oxygen and carbon dioxide concentrations. Data were collected and analyzed with Panlab METABOLISM software (Vídeňská, Prague, Czech Republic).

2.4. Glucose Tolerance Test

Mice were fasted overnight for a total of 16 h (18:00 until 10:00 the next day). At 10:00, blood was drawn via tail vein puncture and baseline glycemia measured using a OneTouch Ultra2 glucometer (Lifescan, Milpitas, CA, USA). Mice were then administered glucose (1 g/kg body weight) via intraperitoneal bolus injection and glycemia measured at 30 min, 1 h, 2 h and 3 h post-injection.

2.5. Tissue Collection and Processing

Prior to tissue collection, final body weight and body length were recorded. For immunohistochemical analysis, mice were euthanized with tribromoethanol (1%, 0.1 mL/g body weight), blood collected via cardiac puncture, and perfused transcardially with phosphate buffer, followed by 4% paraformaldehyde (PFA) in phosphate buffer. Brains were collected in 4% PFA for overnight post-fix followed by dehydration via sucrose gradient, then cut into 40 μm floating coronal sections using a cryostat. Floating sections were stored in a cryoprotective solution (50% 0.1 M phosphate buffer, 25% glycerol, 25% ethylene glycol) until analyzed. Brown adipose tissue (BAT) was collected and stored in 4% PFA for several days before being paraffin-embedded and cut into 5 μm sections. For Western blot analysis of fresh-frozen tissue, mice were euthanized with CO_2, blood collected and inguinal white adipose tissue removed and weighed. Brains were placed in 30% sucrose on ice for 1 min, then the hypothalamus was dissected and snap-frozen in liquid nitrogen.

Frozen hypothalami were pulverized using the CryoGrinder kit (OPS Diagnostics, Lebanon, NJ, USA; CG 08-01). Individual brain parts were placed in the ceramic mortar on dry ice and ground into powder using the ceramic pestle. The powder was added to a tube containing 300 μL CelLytic MT Mammalian Tissue Lysis/Extraction Reagent (Sigma, St. Louis, MO, USA; C-3228) containing a protease/phosphatase inhibitor cocktail (1:100; Cell Signaling, Danvers, MA, USA; 5872) and sonicated with 20 one-second pulses at 5 Hz, separated by one second each, using a Fisher Sonic Dismembrator Model 100 (Fisher Scientific, Hampton, NH, USA). Samples were then centrifuged at 14,000× g for 10 min at 4 °C and supernatant was collected and stored at −80 °C for Western blotting.

2.6. Gel Electrophoresis and Western Blotting

Hypothalamic lysate samples containing 40 μg of protein were separated on 4%–20% gradient polyacrylamide TGX gels (BIO-RAD, Hercules, CA, USA; 5671094) via electrophoresis and transferred to 0.45 μm pore size Immobilon-FL polyvinylidene difluoride membranes (Millipore, Burlington, MA, USA; IPFL00010). Membranes were incubated in PBS-based blocking buffer (LI-COR Biosciences, Lincoln, NE, USA; P/N 927) for 1 h and then probed with primary antibodies overnight at 4 °C with shaking, followed by washing with PBS containing 0.01% Tween 20 (PBS-T). Blots were incubated with infrared fluorophore-bound secondary antibodies in the dark, washed again with PBS-T, and analyzed using the Odyssey CLx Imaging System (LI-COR Biosciences). After the phosphorylated signal transducer and activator of transcription 3 (pSTAT3) was measured, membranes were stripped (Re-Blot Plus Strong Solution; Millipore, Burlington MA, USA; 2504) and then probed for STAT3 to generate a pSTAT3/STAT3 ratio.

2.7. Immunohistochemistry and Histology

For the detection of proteins by 3,3′-diaminobenzidin (DAB), endogenous peroxidases were inactivated with 1% H_2O_2 in methanol. Sections were blocked in normal goat serum and incubated overnight at 4 °C in primary antibody. Next, sections were probed with the appropriate biotinylated secondary antibodies, then incubated with an avidin-biotin-peroxidase complex (Elite ABC Kit; Vector Labs, Burlingame, CA, USA; PK-6100). DAB chromogen (DAB Substrate Kit; Vector Labs; H-2200) was used for peroxidase detection of immunoreactivity. Finally, sections were rinsed with PBS, mounted on glass slides, dehydrated by ethanol gradient followed by xylene, and cover-slipped.

To visualize BAT morphology, 5 μm sections were stained with hematoxylin and eosin (H & E). To evaluate UCP1 levels, sections were placed in a 60 °C oven for 30 min, deparaffinized with xylene and ethanol, and incubated in 1% H_2O_2 in methanol for 30 min. Antigen retrieval was performed with 0.01 M citric acid (pH 6) and sections were blocked using an avidin/biotin blocking kit (Vector Laboratories, Burlingame, CA, USA; SP-2001), followed by incubation with primary antibody and then biotinylated secondary antibody. DAB staining was performed as described above to visualize protein expression and sections were counterstained with hematoxylin before being dehydrated and cover slipped.

For verification of KO of Scly in Agrp neurons, we performed immunofluorescence on a free-floating section from subjects with Agrp-driven expression of the fluorescent dTomato protein. Sections were blocked with normal goat serum, followed by incubation with goat anti-mouse antibody (Jackson Laboratories, West Grove, PA, USA; 115-007-003). Anti-Scly primary antibody was coupled with Alexa Fluor 488 fluorescent secondary antibody (Abcam, Cambridge, MA, USA; ab150113) (1:2 ratio) for 30 min, non-coupled secondary antibody quenched with mouse serum, and the antibody complex was then added to the sections and incubated overnight at 4 °C with shaking. Representative images were captured on a Leica TSP SP8 HyVolution Confocal Microscope (Leica Biosystems, Buffalo Grove, IL, USA).

2.8. Stereology and Data Quantification

Sections at bregma −1.46 mm were used for analysis of the Arc, median eminence (ME), ventromedial hypothalamus (VMH), and dorsomedial hypothalamus (DMH) (illustrated in Supplementary Figure S3). To analyze the paraventricular nucleus (PVN) and periventricular nucleus, sections at bregma −0.94 mm were used. Analysis was performed with Stereo Investigator Software (MBF Bioscience, Williston, VT, USA) on an upright microscope (Axioskop2; Zeiss, Oberkochen, Germany). To quantify Agrp neurons, sections at bregma −2.46 mm were visualized with a 5× objective lens, then the total number of dTomato-positive cells were counted in the arcuate nucleus (Arc) of either the left, right or both hemispheres and the mean was calculated for each subject. To measure the optical density of pSTAT3 and Agrp immunoreactivity, 5× brightfield images were captured and imported into ImageJ software. Images were converted to black-and-white, inverted and the mean value per pixel measured within each region of interest. The ImageJ plugin Cell Counter was used to count pSTAT3-positive cells.

The simple random sampling workflow in Stereo Investigator was used to take up to 10 images of BAT sections stained with H&E. Images were analyzed using the FIJI (ImageJ v2) plugin Adiposoft to count and measure the size of individual lipid droplets, and values were averaged for each subject. To measure UCP1 optical density, images were converted to black-and-white, inverted and the mean pixel value for the entire image was measured. Lipid droplets have the potential to bias data since they comprise a significant percentage of each image and do not express UCP1. To account for this, the mean pixel value for an individual lipid droplet was measured, multiplied by the area of the image filled by lipid deposition, and this value was subtracted from the optical density measured for the entire image to calculate the overall intensity of UCP1 staining.

2.9. Antibodies

The following primary antibodies were used: rabbit anti-mouse Agouti-related protein antiserum (1:1000; Alpha Diagnostic International, Inc., San Antonio, TX, USA; AGRP11-S), rabbit anti-phospho-STAT3 (Tyr705) (D3A7) (1:1000; Cell Signaling, Danvers, MA, USA; 9145), rabbit anti-STAT3 (D1A5) (1:1000; Cell Signaling, Danvers, MA, USA; 8768), mouse anti-selenocysteine lyase (32) (1:200; Santa Cruz Biotechnology, Santa Cruz, CA, USA; sc-136394), rabbit anti-β-actin (13E5) (1:5000; Cell Signaling, Danvers, MA, USA; 4970), rabbit anti-uncoupling protein 1 (1:500; Abcam, Cambridge, MA, USA; ab10983).

2.10. Statistical Analysis

Statistical tests and sample numbers varied with each assay performed and are indicated in the figure legends. Generally, data sets were analyzed by two-way ANOVA to detect changes caused by genotype and either sex or leptin treatment, and graphed accordingly. Tukey's multiple comparisons test was used for post-hoc analysis, unless a repeated measures design analysis was performed, in which case, Bonferroni's multiple comparisons test was used. In cases of immunohistochemical analysis of brain sections, separate two-way ANOVAs were performed for each hypothalamic region to compare genotype and leptin response. Sex-wise statistical comparisons were not performed on DAB-stained brain sections since staining was performed separately. Data were analyzed and plotted using GraphPad Prism version 7 software. All results are represented as mean ± standard error of the mean (SEM). Significance was determined by a p-value of <0.05. Sample sizes, 'n', reported in graphs and figure legends, represent biological replicates. Technical replicates (images from a single section) were sampled from BAT sections, as described in Section 2.8 above, and the mean was calculated for each biological specimen.

3. Results

3.1. Scly-Agrp KO Are Resistant to HFD-Induced Weight Gain

Knockout of Scly was verified via immunofluorescence (Supplementary Materials Figure S1). Scly-Agrp KO mice and littermate controls were placed on HFD beginning at 4 weeks of age and metabolically evaluated thereafter. Surprisingly, both male and female Scly-Agrp KO mice gained less weight while on HFD than controls (Figure 1A,B), and had reduced inguinal fat deposits (Figure 1C). A comparison of the total body weight of both sexes of control and Scly-Agrp KO mice on HFD over time gave the same result (Supplementary Materials Figure S2C,D). Body length and the ratio of body weight to length were also reduced in Scly-Agrp KO mice (Supplementary Materials Figure S2E,F). Despite gaining less weight, Scly-Agrp KO mice did not exhibit any change in food consumption, although there was a downward trend in females during the dark phase (Figure 1D–G). While there were no significant changes in glucose tolerance (Supplementary Materials Figure S2G–I), fasting serum leptin levels trended towards a decrease in Scly-Agrp KO mice, possibly reflecting a change in leptin sensitivity (Figure 1H). Overall, metabolic characterization of Scly-Agrp KO mice revealed an anorexigenic phenotype that is in stark contrast to whole-body Scly KO mice.

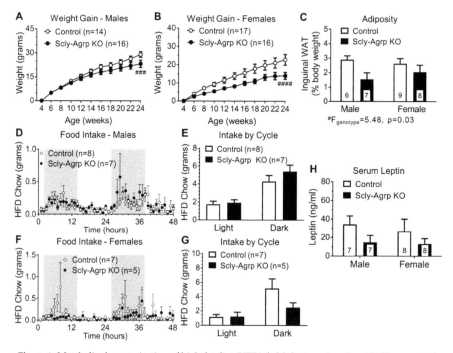

Figure 1. Metabolic characterization of high-fat diet (HFD)-fed Scly-Agrp knockout (KO) vs. control mice. (**A**) Weight gain on a high-fat diet in male Scly-Agrp KO mice vs. controls. Two-way ANOVA: genotype ### $F_{(1,308)}$ = 14.24, p = 0.0002; (**B**) Weight gain in females. Two-way ANOVA: genotype #### $F_{(1,341)}$ = 48.88, p < 0.0001; (**C**) Inguinal fat deposits, expressed as percent body weight, by sex and genotype. Two-way ANOVA: genotype # $F_{(1,26)}$ = 5.48, p = 0.03; (**D**) Food consumption over a 48-h period in male mice. Gray shading indicates dark cycle; (**E**) Comparison of total intake in males by cycle: Two-way ANOVA: genotype $F_{(1,26)}$ = 1.349, p = 0.26; (**F**) Food consumption in female mice; (**G**) Comparison of total intake in females by cycle. Two-way ANOVA: genotype $F_{(1,20)}$ = 2.118, p = 0.16, interaction $F_{(1,20)}$ = 2.303, p = 0.14; (**H**) Fasting serum leptin levels in male and female mice by genotype. Two-way ANOVA: genotype $F_{(1,26)}$ = 3.188, p = 0.086. All data are represented as mean ± standard error of the mean. Group numbers are indicated in each graph.

3.2. Scly-Agrp KO Mice Do Not Exhibit HFD-Induced Leptin Resistance in the Arcuate Nucleus.

High-fat diet consumption promotes leptin resistance in rodents [25,26]. Western blot analysis of whole-hypothalamus lysates from leptin-injected (1 mg/kg body weight) control mice fed a HFD did not reveal a significant change in phosphorylation of the leptin signaling protein signal transducer and activator of transcription (pSTAT3), indicating that the mice had developed leptin resistance. Scly-Agrp KO mice fed a HFD, however, exhibited an increase in pSTAT3 in response to leptin, demonstrating that leptin sensitivity was maintained in this group (Figure 2A,B). Immunohistochemical interrogation of hypothalamic regions (see Supplementary Materials Figure S3 for anatomical delineation) revealed that control mice developed leptin resistance in the arcuate nucleus (Arc) and median eminence (ME), which contain the Agrp neuron cell bodies (Figure 2C–E). The dorsomedial hypothalamus (DMH) and ventromedial hypothalamus (VMH), both leptin-responsive regions that do not contain Agrp neurons, maintained leptin sensitivity. Scly-Agrp KO mice, on the other hand, exhibited a robust increase in pSTAT3 in all regions, including the Arc, in response to leptin (Figure 2C–E). This same effect was observed whether measuring pSTAT3 optical density or counting pSTAT3-positive cells (Supplementary Materials Figure S4). These results indicate that while HFD induced Arc- and

ME-specific leptin resistance in control mice, Scly-Agrp KO mice were protected from developing leptin resistance.

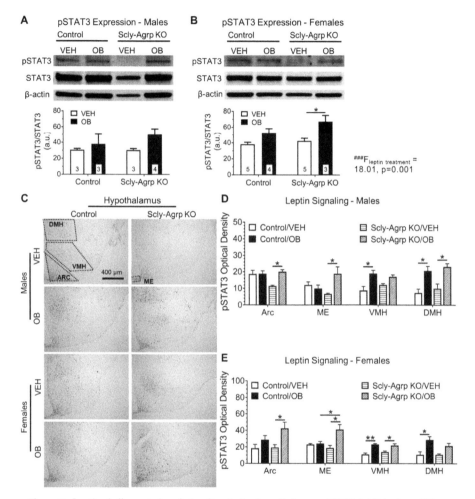

Figure 2. Leptin challenge-induced signaling in the hypothalamus of HFD-fed Scly-Agrp KO vs. control mice. (**A**) Western blot analysis of phosphorylated STAT3 (pSTAT3) levels in hypothalamic protein lysates from male mice, following intraperitoneal (i.p.) injection of leptin (Ob, 1 mg/kg body weight) or vehicle control (VEH, phosphate-buffered saline). Two-way ANOVA: leptin treatment $F_{(1,9)} = 3.26$, $p = 0.1$; (**B**) pSTAT3 levels in female hypothalamic protein lysates. Two-way ANOVA: leptin treatment [###] $F_{(1,13)} = 18.01$, $p = 0.001$; (**C**) Sample images of hypothalamic sections stained for pSTAT3 at 10× magnification; (**D**) Optical density of pSTAT3 measured in male mice in the arcuate nucleus (Arc): Two-way ANOVA: leptin $F_{(1,9)} = 5.889$, $p = 0.038$, interaction $F_{(1,9)} = 5.537$ $p = 0.043$, median eminence (ME): interaction $F_{(1,9)} = 8.409$, $p = 0.018$, ventromedial hypothalamus (VMH): leptin $F_{(1,10)} = 15.02$, $p = 0.003$, and dorsomedial hypothalamus (DMH): leptin $F_{(1,8)} = 27.9$, $p = 0.0007$; (**E**) pSTAT3 optical density in female mouse; Arc: Two-way ANOVA: leptin $F_{(1,10)} = 8.777$, $p = 0.014$, ME: leptin $F_{(1,8)} = 10.35$, $p = 0.012$, interaction $F_{(1,8)} = 8.4$ $p = 0.02$, VMH: leptin $F_{(1,10)} = 36.54$, $p = 0.0001$, and DMH: leptin $F_{(1,8)} = 19.96$, $p = 0.002$. All data are represented as mean ± standard error of the mean. Group numbers are indicated in graphs. Group numbers for (**D,E**) ranged from 3–5. Tukey's multiple comparisons test: * $p < 0.05$, ** $p < 0.01$.

3.3. Scly-Agrp KO Mice Have Less Agrp Neurons and Reduced Hypothalamic Agrp Immunoreactivity Compared to Controls

The under-weight phenotype displayed by Scly-Agrp KO mice while on HFD suggests that the overall influence of Agrp neurons may be reduced. The Scly-Agrp KO mice also have Agrp-Cre-driven expression of a dTomato reporter gene. The number of dTomato-positive cells trended towards a reduction in hypothalamic sections from Scly-Agrp KO mice, suggesting the presence of fewer Agrp neurons (Figure 3A,B). Total hypothalamic Agrp expression measured via Western blot was significantly reduced in female Scly-Agrp KO mice (Figure 3C,D) while trending downward in males. Thus, Scly-Agrp KO mice appear to have less Agrp-ergic activity, which likely drives the metabolic phenotype observed.

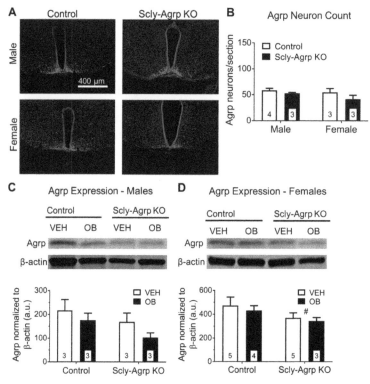

Figure 3. Agrp neuron count and hypothalamic Agrp protein expression in HFD-fed Scly-Agrp KO vs control mice. (**A**) Sample images of hypothalamic sections with dTomato-positive (red) Agrp neurons, in Scly-Agrp KO mice and Scly$^{WT;Agrp-Cre}$ mice used as controls, counter-stained with 4′6-diamidino-2-phenyllindole (DAPI; blue), at 10× magnification; (**B**) Agrp neuron counts trended towards a reduction in male and female Scly-Agrp KO mice. Two-way ANOVA: genotype # $F_{(1,9)} = 3.016$, $p = 0.12$; (**C**) Agrp expression in hypothalamic protein lysates from males injected with vehicle (VEH) or leptin (OB), measured via Western blot and expressed as arbitrary units (a.u.). Two-way ANOVA: genotype $F_{(1,8)} = 4.73$, $p = 0.12$, leptin treatment $F_{(1,8)} = 2.41$, $p = 0.16$; (**D**) Agrp expression in females. Two-way ANOVA: genotype # $F_{(1,8)} = 3.03$, $p = 0.049$. All data are represented as mean ± standard error of the mean. Group numbers are indicated in each graph. # denotes a significant genotype effect.

Agrp neurons project to multiple regions of the hypothalamus, including other cells within the Arc, to affect the energy balance [22]. Therefore, we measured Agrp expression in these areas to determine whether Agrp innervation was reduced within a particular neurological circuit (see Supplementary Materials Figure S3 for anatomical delineation). Agrp immunoreactivity was found to be reduced in the Arc and DMH of both male and female Scly-Agrp KO mice (Figure 4A–C). This strongly implies that Scly-Agrp KO mice have reduced Agrp circuitry to these specific hypothalamic regions, which may mediate the underweight phenotype of Scly-Agrp KO mice.

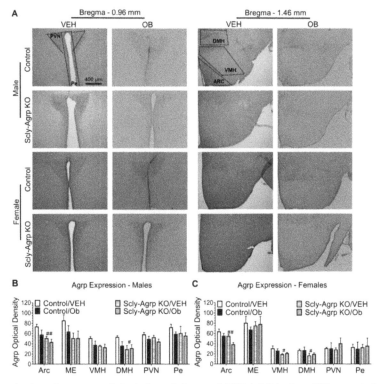

Figure 4. Agrp immunoreactivity in hypothalamus of HFD-fed Scly-Agrp KO vs control mice. (**A**) Sample images of hypothalamic sections showing Agrp immunoreactivity at 10× magnification; (**B**) Agrp expression in hypothalamic sections from male mice, injected with either vehicle (VEH) or leptin (OB), analyzed via two-way ANOVA in the arcuate nucleus (Arc): [##] genotype $F_{(1,10)} = 10.54$, $p = 0.009$, leptin treatment $F_{(1,10)} = 4.11$, $p = 0.07$, median eminence (ME), ventromedial hypothalamus (VMH), dorsomedial hypothalamus (DMH): [#] genotype $F_{(1,9)} = 1.541$, $p = 0.043$, paraventricular nucleus (PVN): leptin treatment $F_{(1,10)} = 4.04$, $p = 0.072$, and periventricular nucleus (Pe); (**C**) Agrp expression in female mice by hypothalamic region: Two-way ANOVA: Arc: [##] genotype $F_{(1,10)} = 11.9$, $p = 0.006$, leptin treatment $F_{(1,10)} = 11.36$, $p = 0.007$, VMH: [#] genotype $F_{(1,10)} = 8.87$, $p = 0.014$, DMH: [#] genotype $F_{(1,10)} = 5.38$, $p = 0.04$. All data are represented as mean ± standard error of the mean. Group numbers ranged from 3–5. # denotes a significant genotype effect within a particular brain region.

Leptin regulates Agrp production in the mouse hypothalamus [27]. Interestingly, two-way ANOVA revealed a significant effect of leptin treatment on Agrp expression in the Arc of female mice (Figure 4C). Post-hoc analysis revealed a significant reduction in Agrp expression in response to leptin in the Arc of female Scly-Agrp KO mice, but not in control mice. Leptin had a similar effect in male mice, although the results were not statistically significant (Figure 4B). Considering that Scly-Agrp KO mice maintained leptin sensitivity on HFD while control mice developed leptin resistance (Figure 2), this result provides further insight into a potential mechanism involving leptin regulation of Agrp activity in Scly-Agrp KO mice.

3.4. Brown Adipose Tissue Morphology Is Altered in Scly-Agrp KO Mice.

Although Agrp neurons are primarily known to influence energy metabolism by promoting feeding behavior, there is growing evidence for a prominent role of Agrp neurons in regulating brown adipose tissue (BAT) thermogenesis [28–30]. Since the Scly-Agrp KO mice did not exhibit any changes in food intake, we investigated BAT for changes in morphology and thermogenesis. Analysis of H&E-stained BAT sections revealed that lipid droplets in HFD-fed Scly-Agrp KO mice were significantly smaller than in HFD-fed controls, suggesting elevated thermogenesis (Figure 5A–C). While this effect was more pronounced in the male mice, the female mice generally had smaller lipid droplets to begin with, which is consistent with past literature [31]. Overall fat deposition, measured as the fraction of space occupied by lipid mass, was also reduced in Scly-Agrp KO mice (Supplementary Materials Figure S5).

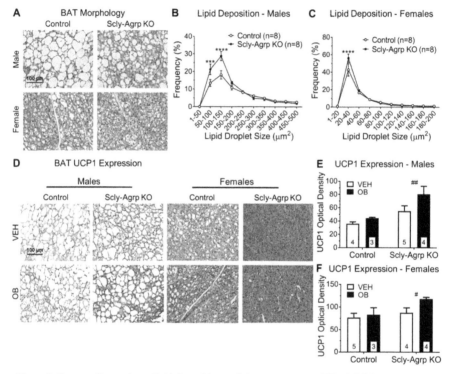

Figure 5. Brown adipose tissue lipid deposition and thermogenesis in HFD-fed Scly-Agrp KO vs control mice. (**A**) Sample images of brown adipose tissue (BAT) sections stained with hematoxylin and eosin, showing adipocytes and lipid droplets at 40× magnification; (**B**) frequency distribution of lipid droplet size in BAT sections from male mice. Two-way ANOVA with repeated measures: genotype $F_{(1,14)} = 8.079$, $p = 0.013$, interaction $F_{(9,126)} = 5.726$, $p < 0.0001$, Bonferroni's multiple comparisons test: *** $p < 0.001$, **** $p < 0.0001$; (**C**) BAT lipid droplet size in female mice. Two-way ANOVA with repeated measures: interaction $F_{(9,126)} = 2.418$, $p = 0.014$, Bonferroni's multiple comparisons test: **** $p < 0.0001$. Treatment with either vehicle (VEH) or leptin (OB) did not affect lipid droplet size; (**D**) sample images of uncoupling protein-1 (UCP1) immunoreactivity in BAT sections counter-stained with hematoxylin; (**E**) UCP1 expression in male BAT. Two-way ANOVA: ## genotype $F_{(1,12)} = 11.6$, $p = 0.005$, leptin treatment $F_{(1,12)} = 4.35$, $p = 0.06$; interaction $F_{(1,12)} = 1.407$, $p = 0.26$ (**F**) UCP1 in female BAT. Two-way ANOVA: # genotype $F_{(1,12)} = 5.0$, $p = 0.04$, leptin treatment $F_{(1,12)} = 3.27$, $p = 0.09$, interaction $F_{(1,12)} = 1.35$, $p = 0.3$. All data are represented as mean ± standard error of the mean. Group numbers are indicated in each graph. # denotes a significant genotype effect.

Mitochondrial uncoupling protein 1 (UCP1) activates BAT thermogenesis, which promotes energy expenditure and fat loss [32]. BAT sections from Scly-Agrp KO mice had significantly greater amounts of UCP1 immunoreactivity than BAT sections from control mice, suggesting elevated thermogenic activity in Scly-Agrp KO mice (Figure 5D,E). These data implicate BAT thermogenesis as a contributing factor in the anorexigenic effect of Agrp neuron-specific KO of Scly.

4. Discussion

This study was initiated with the hypothesis that knocking out Scly in Agrp neurons would re-capitulate the obesogenic phenotype of whole-body Scly KO mice [13]. Unexpectedly, Scly-Agrp KO mice gained less weight and adiposity than controls while on HFD (Figure 1), demonstrating the complexity of the role of Scly in hypothalamic function. Previous work on whole-body Scly KO mice revealed decreased expression of multiple selenoproteins in the hypothalamus, including glutathione peroxidase 1 (GPx1), which breaks down hydroperoxides, thus reducing oxidative stress, as well as SelM and selenoprotein S (SelS), both of which mitigate ER stress [14]. The role of ER stress in causing leptin resistance has been established in previous studies using HFD-fed rodents [25,26] and, recently, SelM has been implicated as a mediator of leptin signaling [19]. However, knocking out Scly in Agrp neurons did not cause leptin resistance in our study, and instead conferred protection from Arc- and ME-specific leptin resistance (Figure 2). Previous work by Diano et al., however, demonstrated the capacity of reactive oxygen species (ROS) to suppress Agrp activity [33]. In this study, intracerebroventricular injection of the ROS scavenger honokiol increased Agrp neuron activity, while ROS induction via administration of GW99662, antagonist of peroxisome proliferator-activated receptor-γ (PPAR-γ), suppressed the firing rate of Agrp neurons. Thus, it is possible that the loss of selenoproteins that limit ROS levels, such as GPx1, permitted an overall reduction in Agrp neuron activity in Scly-Agrp KO mice that outweighed any reduction in Agrp neuron leptin sensitivity.

Scly-Agrp KO mice may have had fewer Agrp neurons than controls and less Agrp expression throughout the hypothalamus (Figures 3 and 4), which could contribute to the overall anorexigenic phenotype. One explanation is that oxidative stress induced by the loss of Scly led to the progressive degeneration of some Agrp neurons. Under-weight phenotypes similar to what we observed in the Scly-Agrp KO mice have been reported in other studies in which progressive degeneration of Agrp neurons was induced. For example, while studying a mouse model in which Agrp neurons were progressively ablated by deleting the mitochondrial transcription factor A (Tfam), Xu et al. observed that the mutant mice had reduced body weight and adiposity [34]. Subsequent investigation of the same mouse model by Pierce and Xu revealed that, in response to Agrp neuron degeneration, the hypothalamus generated new cells, some of which became Agrp neurons [35]. Interestingly, this hypothalamic de novo neurogenesis described by the authors gave rise to leptin-responsive cells. It is, therefore, possible that a similar mechanism involving the generation of new leptin-sensitive neurons to replace degenerating Agrp neurons may have contributed to the apparent resistance of Scly-Agrp KO mice to Arc- and ME-specific leptin resistance.

Although Scly-Agrp KO mice gained less weight and had reduced adiposity compared to control mice, no changes in food intake were observed (Figure 1). This is consistent with the feeding behavior of whole-body Scly KO mice which, although they display hyperphagia when placed on a Se-deficient diet [9], do not consume more food than controls while on HFD [13]. Agrp neurons are typically described as promoting feeding behavior as part of the melanocortin system by inhibiting melanocortin-4 receptor (MC4R)-positive neurons in the PVN of the hypothalamus to suppress anorexigenic hormone signals targeting the pituitary gland. However, in addition to observing no change in food intake in Scly-Agrp KO mice, we also found that Agrp immunoreactivity in the PVN was similar between controls and KO mice. Although an effect on the melanocortin system cannot be ruled out, Agrp neurons project to other hypothalamic regions that may also contribute to the anorexigenic phenotype observed. Immunohistochemical analysis showed decreased Agrp expression in one such area, the DMH (Figure 4), which is a known regulator of BAT thermogenesis [36]. Glutamatergic neurons in the DMH project onto sympathetic premotor neurons in the rostral raphe pallidus (rRPa) which subsequently promote BAT thermogenesis via brain stem circuitry [37–39]. Agrp neurons provide tonic inhibitory input to the DMH, which limits sympathetic nerve activity [40]. Disinhibiting DMH glutamatergic neurons, as would be the case with decreased input from Agrp neurons, has been shown to stimulate BAT thermogenesis [41]. Interestingly, Scly-Agrp KO mouse BAT sections had smaller amounts of lipid deposition and increased UCP1 expression, suggesting elevated levels of thermogenesis (Figure 5). Together these data implicate enhanced BAT thermogenesis, possibly involving the DMH-sympathetic pathway, as an underlying cause of the resistance to HFD-induced weight gain exhibited by Scly-Agrp KO mice.

The results discussed in this report elucidate a previously undescribed role of Scly in regulating body composition via the hypothalamus. We have shown that loss of Scly within a neuronal sub-population that makes up a small fraction of the hypothalamus [22] produces a phenotype in mice distinguished by reduced weight gain while on HFD and resistance to the development of leptin insensitivity. Moreover, our data reveal yet another way that Scly can influence energy homeostasis: via Agrp neuron-mediated BAT activation. Overall, these findings provide novel insights into the importance of Se utilization in central nervous system-directed energy metabolism.

Supplementary Materials: The following are available online at http://www.mdpi.com/2072-6643/11/7/1693/s1, Figure S1: Verification of Scly KO in Agrp neurons. Immunofluorescent labeling of Scly in sections of arcuate nucleus in control and Scly-Agrp KO mice taken at 40× and counter-stained with 4'6-diamidino-2-phenyllindole (DAPI; blue). Agrp neurons are positive for dTomato fluorescent protein expression (red) and are labeled with orange arrows. Agrp neurons are positive for Scly (green) in sections from control mice, whereas Agrp neurons in sections from Scly-Agrp KO mice are not positive for Scly, Figure S2: Comparison of weight gain in Scly$^{fl/fl;\,Agrp-WT}$ vs Scly$^{WT;\,Agrp-Cre}$ mice and comparison of body length, total body weight and glucose tolerance in Scly-Agrp KO vs control mice. (**a**) Change in total body weight in male Scly-Agrp KO mice vs controls. Two-way ANOVA: genotype $F_{(1,308)} = 50.24$, $p < 0.0001$; (**b**) Body weight change in females. Two-way ANOVA: genotype $F_{(1,341)} = 40.44$, $p < 0.0001$; (**c**) Scly$^{fl/fl;\,Agrp-WT}$ vs Scly$^{WT;\,Agrp-Cre}$ weight gain in males. Two-way ANOVA: genotype $F_{(1,253)} = 3.866$, $p = 0.05$; (**d**) and females: genotype $F_{(1,307)} = 5.241$, $p = 0.02$; (**e**) Body lengths of Scly-Agrp KO vs control mice. Two-way ANOVA: genotype $F_{(1,53)} = 8.093$, $p = 0.006$; (**f**) Body weight/body length. Two-way ANOVA: genotype $F_{(1,48)} = 9.635$, $p = 0.003$; (**g**) Time-course of blood glycemia following i.p. bolus injection of glucose (1g/kg body weight) in males; (**h**) and females; (**i**) Comparison of area under the curve representing cumulative glycemia scores over a 3-h period post-injection. Two-way ANOVA: genotype $F_{(1,45)} = 0.1589$, $p = 0.69$, sex $F_{(1,45)} = 8.107$, $p = 0.007$. All data are represented as mean ± standard error of the mean. Group numbers are indicated in each graph. Tukey's multiple comparisons test: * $p < 0.05$. Figure S3: Representative images of hypothalamic sections stained for Agrp, at 5X magnification, to show brain regions. (**a**) Bregma −0.94mm hypothalamus containing paraventricular nucleus (PVN) and periventricular nucleus (Pe); (**b**) Bregma −1.46mm containing arcuate nucleus (Arc), median eminence (ME), ventromedial hypothalamus (VMH), and dorsomedial hypothalamus (DMH), Figure S4: Number of phosphorylated STAT3 (pSTAT3)-positive cells counted in response to leptin (Ob, 1mg/kg body weight) or vehicle (VEH, phosphate-buffered saline) injection in Scly-Agrp KO vs control mice. (**a**) Cell counts of pSTAT3-positive cells measured in male mouse arcuate nucleus (Arc): Two-way ANOVA: leptin $F_{(1,9)} = 11.47$, $p = 0.008$, genotype $F_{(1,9)} = 7.06$ $p = 0.026$, median eminence (ME): genotype $F_{(1,9)} = 4.421$ $p = 0.065$, ventromedial hypothalamus (VMH): leptin $F_{(1,10)} = 17.29$, $p = 0.002$, and dorsomedial hypothalamus (DMH): leptin $F_{(1,9)} = 9.126$, $p = 0.015$; (**b**) pSTAT3-cells in female mouse Arc: Two-way ANOVA: leptin $F_{(1,10)} = 7.318$, $p = 0.02$, ME: leptin $F_{(1,8)} = 12.3$, $p = 0.008$, VMH: leptin $F_{(1,10)} = 31.81$, $p = 0.0002$, genotype $F_{(1,10)} = 16.42$, $p = 0.002$, interaction $F_{(1,10)} = 5.543$, p = 0.04 and DMH: leptin $F_{(1,8)} = 29.8$, $p = 0.0006$, genotype $F_{(1,8)} = 6.951$ $p = 0.03$. All data are represented as mean ± standard error of the mean. Group numbers are indicated in each graph. Tukey's multiple comparisons test: * $p < 0.05$, ** $p < 0.01$, Figure S5: Brown adipose tissue (BAT) lipid fraction represented as the percent of surface area of the section occupied by lipid droplet. Two-way ANOVA: genotype $F(1,28) = 6.474$, $p = 0.017$, sex $F(1,28) = 4.894$, $p = 0.035$. All data are represented as mean ± standard error of the mean. Group numbers are indicated in each graph.

Author Contributions: Conceptualization, D.J.T., M.W.P. and M.J.B.; Data curation, D.J.T. and A.C.H.; Formal analysis, D.J.T.; Funding acquisition, M.J.B.; Investigation, D.J.T.; Methodology, M.W.P. and A.C.H.; Project administration, M.J.B.; Resources, A.C.H. and M.J.B.; Supervision, M.W.P.; Validation, D.J.T.; Writing—original draft, D.J.T.; Writing—review & editing, M.W.P. and M.J.B.

Funding: This research was funded by The National Institutes of Health, grant numbers R01DK047320 (M.J.B.) and U54MD007601 which supported core facilities, and a Research Supplements to Promote Diversity in Health-Related Research, R01DK047320-22S2 (D.J.T.).

Acknowledgments: The authors are grateful to Lucia Seale for providing reagents and for insightful input on experimental design and data interpretation.

Conflicts of Interest: The authors declare no conflict of interest. The funders had no role in the design of the study; in the collection, analyses, or interpretation of data; in the writing of the manuscript, or in the decision to publish the results.

References

1. Schomburg, L. Selenium, selenoproteins and the thyroid gland: interactions in health and disease. *Nat. Rev. Endocrinol.* **2011**, *8*, 160–171. [CrossRef]

2. Flohe, L. Selenium in mammalian spermiogenesis. *Biol. Chem.* **2007**, *388*, 987–995. [CrossRef] [PubMed]

3. Youn, H.S.; Lim, H.J.; Choi, Y.J.; Lee, J.Y.; Lee, M.Y.; Ryu, J.H. Selenium suppresses the activation of transcription factor NF-kappa B and IRF3 induced by TLR3 or TLR4 agonists. *Int. Immunopharmacol.* **2008**, *8*, 495–501. [CrossRef] [PubMed]

4. Ogawa-Wong, A.N.; Berry, M.J.; Seale, L.A. Selenium and Metabolic Disorders: An Emphasis on Type 2 Diabetes Risk. *Nutrients* **2016**, *8*, 80. [CrossRef] [PubMed]

5. Loh, K.; Deng, H.; Fukushima, A.; Cai, X.; Boivin, B.; Galic, S.; Bruce, C.; Shields, B.J.; Skiba, B.; Ooms, L.M.; et al. Reactive oxygen species enhance insulin sensitivity. *Cell Metab.* **2009**, *10*, 260–272. [CrossRef]

6. Misu, H.; Takamura, T.; Takayama, H.; Hayashi, H.; Matsuzawa-Nagata, N.; Kurita, S.; Ishikura, K.; Ando, H.; Takeshita, Y.; Ota, T.; et al. A liver-derived secretory protein, selenoprotein P, causes insulin resistance. *Cell Metab.* **2010**, *12*, 483–495. [CrossRef]

7. Pitts, M.W.; Reeves, M.A.; Hashimoto, A.C.; Ogawa, A.; Kremer, P.; Seale, L.A.; Berry, M.J. Deletion of selenoprotein M leads to obesity without cognitive deficits. *J. Biol. Chem.* **2013**, *288*, 26121–26134. [CrossRef] [PubMed]

8. De Jesus, L.A.; Carvalho, S.D.; Ribeiro, M.O.; Schneider, M.; Kim, S.W.; Harney, J.W.; Larsen, P.R.; Bianco, A.C. The type 2 iodothyronine deiodinase is essential for adaptive thermogenesis in brown adipose tissue. *J. Clin. Investig.* **2001**, *108*, 1379–1385. [CrossRef]

9. Seale, L.A.; Hashimoto, A.C.; Kurokawa, S.; Gilman, C.L.; Seyedali, A.; Bellinger, F.P.; Raman, A.V.; Berry, M.J. Disruption of the selenocysteine lyase-mediated selenium recycling pathway leads to metabolic syndrome in mice. *Mol. Cell Biol.* **2012**, *32*, 4141–4154. [CrossRef]

10. Ha, H.Y.; Alfulaij, N.; Berry, M.J.; Seale, L.A. From Selenium Absorption to Selenoprotein Degradation. *Biol. Trace Elem. Res.* **2019**. [CrossRef]

11. Kurokawa, S.; Takehashi, M.; Tanaka, H.; Mihara, H.; Kurihara, T.; Tanaka, S.; Hill, K.; Burk, R.; Esaki, N. Mammalian selenocysteine lyase is involved in selenoprotein biosynthesis. *J. Nutr. Sci. Vitaminol.* **2011**, *57*, 298–305. [CrossRef] [PubMed]

12. Mihara, H.; Kurihara, T.; Watanabe, T.; Yoshimura, T.; Esaki, N. cDNA cloning, purification, and characterization of mouse liver selenocysteine lyase. Candidate for selenium delivery protein in selenoprotein synthesis. *J. Biol. Chem.* **2000**, *275*, 6195–6200. [CrossRef] [PubMed]

13. Seale, L.A.; Gilman, C.L.; Hashimoto, A.C.; Ogawa-Wong, A.N.; Berry, M.J. Diet-induced obesity in the selenocysteine lyase knockout mouse. *Antioxid. Redox Signal.* **2015**, *23*, 761–774. [CrossRef] [PubMed]

14. Ogawa-Wong, A.N.; Hashimoto, A.C.; Ha, H.; Pitts, M.W.; Seale, L.A.; Berry, M.J. Sexual Dimorphism in the Selenocysteine Lyase Knockout Mouse. *Nutrients* **2018**, *10*, 159. [CrossRef] [PubMed]

15. Thaler, J.P.; Yi, C.X.; Schur, E.A.; Guyenet, S.J.; Hwang, B.H.; Dietrich, M.O.; Zhao, X.; Sarruf, D.A.; Izgur, V.; Maravilla, K.R.; et al. Obesity is associated with hypothalamic injury in rodents and humans. *J. Clin. Investig.* **2012**, *122*, 153–162. [CrossRef] [PubMed]

16. Chen, J.; Berry, M.J. Selenium and selenoproteins in the brain and brain diseases. *J. Neurochem.* **2003**, *86*, 1–12. [CrossRef] [PubMed]

17. Yagishita, Y.; Uruno, A.; Fukutomi, T.; Saito, R.; Saigusa, D.; Pi, J.; Fukamizu, A.; Sugiyama, F.; Takahashi, S.; Yamamoto, M. Nrf2 Improves Leptin and Insulin Resistance Provoked by Hypothalamic Oxidative Stress. *Cell Rep.* **2017**, *18*, 2030–2044. [CrossRef]

18. Schriever, S.C.; Zimprich, A.; Pfuhlmann, K.; Baumann, P.; Giesert, F.; Klaus, V.; Kabra, D.G.; Hafen, U.; Romanov, A.; Tschop, M.H.; et al. Alterations in neuronal control of body weight and anxiety behavior by glutathione peroxidase 4 deficiency. *Neuroscience* **2017**, *357*, 241–254. [CrossRef]

19. Gong, T.; Hashimoto, A.C.; Sasuclark, A.R.; Khadka, V.S.; Gurary, A.; Pitts, M.W. Selenoprotein M Promotes Hypothalamic Leptin Signaling and Thioredoxin Antioxidant Activity. *Antioxid. Redox Signal.* **2019**. [CrossRef]

20. Cui, H.; Lopez, M.; Rahmouni, K. The cellular and molecular bases of leptin and ghrelin resistance in obesity. *Nat. Rev. Endocrinol.* **2017**, *13*, 338–351. [CrossRef]

21. Gong, T.; Torres, D.J.; Berry, M.J.; Pitts, M.W. Hypothalamic redox balance and leptin signaling-Emerging role of selenoproteins. *Free Radic. Biol. Med.* **2018**, *127*, 172–181. [CrossRef] [PubMed]

22. Cansell, C.; Denis, R.G.; Joly-Amado, A.; Castel, J.; Luquet, S. Arcuate AgRP neurons and the regulation of energy balance. *Front. Endocrinol.* **2012**, *3*, 169. [CrossRef] [PubMed]

23. Olofsson, L.E.; Unger, E.K.; Cheung, C.C.; Xu, A.W. Modulation of AgRP-neuronal function by SOCS3 as an initiating event in diet-induced hypothalamic leptin resistance. *Proc. Natl. Acad. Sci. USA* **2013**, *110*, E697–E706. [CrossRef] [PubMed]

24. Tong, Q.; Ye, C.P.; Jones, J.E.; Elmquist, J.K.; Lowell, B.B. Synaptic release of GABA by AgRP neurons is required for normal regulation of energy balance. *Nat. Neurosci.* **2008**, *11*, 998–1000. [CrossRef] [PubMed]

25. Zhang, X.; Zhang, G.; Zhang, H.; Karin, M.; Bai, H.; Cai, D. Hypothalamic IKKbeta/NF-kappaB and ER stress link overnutrition to energy imbalance and obesity. *Cell* **2008**, *135*, 61–73. [CrossRef]

26. Ozcan, L.; Ergin, A.S.; Lu, A.; Chung, J.; Sarkar, S.; Nie, D.; Myers, M.G., Jr.; Ozcan, U. Endoplasmic reticulum stress plays a central role in development of leptin resistance. *Cell Metab.* **2009**, *9*, 35–51. [CrossRef] [PubMed]

27. Ebihara, K.; Ogawa, Y.; Katsuura, G.; Numata, Y.; Masuzaki, H.; Satoh, N.; Tamaki, M.; Yoshioka, T.; Hayase, M.; Matsuoka, N.; et al. Involvement of agouti-related protein, an endogenous antagonist of hypothalamic melanocortin receptor, in leptin action. *Diabetes* **1999**, *48*, 2028–2033. [CrossRef] [PubMed]

28. Ruan, H.B.; Dietrich, M.O.; Liu, Z.W.; Zimmer, M.R.; Li, M.D.; Singh, J.P.; Zhang, K.; Yin, R.; Wu, J.; Horvath, T.L.; et al. O-GlcNAc transferase enables AgRP neurons to suppress browning of white fat. *Cell* **2014**, *159*, 306–317. [CrossRef]

29. Burke, L.K.; Darwish, T.; Cavanaugh, A.R.; Virtue, S.; Roth, E.; Morro, J.; Liu, S.M.; Xia, J.; Dalley, J.W.; Burling, K.; et al. mTORC1 in AGRP neurons integrates exteroceptive and interoceptive food-related cues in the modulation of adaptive energy expenditure in mice. *Elife* **2017**, *6*, 22848. [CrossRef] [PubMed]

30. Deng, J.; Yuan, F.; Guo, Y.; Xiao, Y.; Niu, Y.; Deng, Y.; Han, X.; Guan, Y.; Chen, S.; Guo, F. Deletion of ATF4 in AgRP Neurons Promotes Fat Loss Mainly via Increasing Energy Expenditure. *Diabetes* **2017**, *66*, 640–650. [CrossRef] [PubMed]

31. Harshaw, C.; Culligan, J.J.; Alberts, J.R. Sex differences in thermogenesis structure behavior and contact within huddles of infant mice. *PLoS ONE* **2014**, *9*, e87405. [CrossRef] [PubMed]

32. Fedorenko, A.; Lishko, P.V.; Kirichok, Y. Mechanism of fatty-acid-dependent UCP1 uncoupling in brown fat mitochondria. *Cell* **2012**, *151*, 400–413. [CrossRef] [PubMed]

33. Diano, S.; Liu, Z.W.; Jeong, J.K.; Dietrich, M.O.; Ruan, H.B.; Kim, E.; Suyama, S.; Kelly, K.; Gyengesi, E.; Arbiser, J.L.; et al. Peroxisome proliferation-associated control of reactive oxygen species sets melanocortin tone and feeding in diet-induced obesity. *Nat. Med.* **2011**, *17*, 1121–1127. [CrossRef] [PubMed]

34. Xu, A.W.; Kaelin, C.B.; Morton, G.J.; Ogimoto, K.; Stanhope, K.; Graham, J.; Baskin, D.G.; Havel, P.; Schwartz, M.W.; Barsh, G.S. Effects of hypothalamic neurodegeneration on energy balance. *PLoS Biol.* **2005**, *3*, e415. [CrossRef] [PubMed]

35. Pierce, A.A.; Xu, A.W. De novo neurogenesis in adult hypothalamus as a compensatory mechanism to regulate energy balance. *J. Neurosci.* **2010**, *30*, 723–730. [CrossRef] [PubMed]

36. Morrison, S.F.; Nakamura, K.; Madden, C.J. Central control of thermogenesis in mammals. *Exp. Physiol.* **2008**, *93*, 773–797. [CrossRef] [PubMed]

37. Bamshad, M.; Song, C.K.; Bartness, T.J. CNS origins of the sympathetic nervous system outflow to brown adipose tissue. *Am. J. Physiol.* **1999**, *276*, R1569–R1578. [CrossRef] [PubMed]

38. Lee, S.J.; Kirigiti, M.; Lindsley, S.R.; Loche, A.; Madden, C.J.; Morrison, S.F.; Smith, M.S.; Grove, K.L. Efferent projections of neuropeptide Y-expressing neurons of the dorsomedial hypothalamus in chronic hyperphagic models. *J. Comp. Neurol.* **2013**, *521*, 1891–1914. [CrossRef] [PubMed]

39. Oldfield, B.J.; Giles, M.E.; Watson, A.; Anderson, C.; Colvill, L.M.; McKinley, M.J. The neurochemical characterisation of hypothalamic pathways projecting polysynaptically to brown adipose tissue in the rat. *Neuroscience* **2002**, *110*, 515–526. [CrossRef]

40. Shi, Z.; Madden, C.J.; Brooks, V.L. Arcuate neuropeptide Y inhibits sympathetic nerve activity via multiple neuropathways. *J. Clin. Investig.* **2017**, *127*, 2868–2880. [CrossRef]

41. Zaretskaia, M.V.; Zaretsky, D.V.; Shekhar, A.; DiMicco, J.A. Chemical stimulation of the dorsomedial hypothalamus evokes non-shivering thermogenesis in anesthetized rats. *Brain Res.* **2002**, *928*, 113–125. [CrossRef]

Article

Gender Differences with Dose–Response Relationship between Serum Selenium Levels and Metabolic Syndrome—A Case-Control Study

Chia-Wen Lu [1,2], Hao-Hsiang Chang [1], Kuen-Cheh Yang [1,3], Chien-Hsieh Chiang [1,2], Chien-An Yao [1,2] and Kuo-Chin Huang [1,2,3,*]

[1] Department of Family Medicine, National Taiwan University Hospital, Taipei 10002, Taiwan; biopsycosocial@gmail.com (C.-W.L.); allanchanghs@gmail.com (H.-H.C.); quintino.yang@gmail.com (K.-C.Y.); jiansie@ntu.edu.tw (C.-H.C.); yao6638@gmail.com (C.-A.Y.)
[2] Department of Family Medicine, College of Medicine, National Taiwan University, Taipei 10051, Taiwan
[3] Department of Family Medicine, National Taiwan University Hospital Bei-Hu Branch, Taipei 10800, Taiwan
* Correspondence: bretthuang@ntu.edu.tw; Tel: 886-2-23123456 (ext. 66081); Fax: 886-2-2311-8674

Received: 9 January 2019; Accepted: 20 February 2019; Published: 24 February 2019

Abstract: Few studies have investigated the association between selenium and metabolic syndrome. This study aimed to explore the associations between the serum selenium level and metabolic syndrome as well as examining each metabolic factor. In this case-control study, the participants were 1165 adults aged ≥40 (65.8 ± 10.0) years. Serum selenium was measured by inductively coupled plasma-mass spectrometry. The associations between serum selenium and metabolic syndrome were examined by multivariate logistic regression analyses. The least square means were computed by general linear models to compare the serum selenium levels in relation to the number of metabolic factors. The mean serum selenium concentration was 96.34 ± 25.90 µg/L, and it was positively correlated with waist circumference, systolic blood pressure, triglycerides, fasting glucose, and homeostatic model assessment insulin resistance (HOMA-IR) in women, but it was only correlated with fasting glucose and HOMA-IR in men. After adjustment, the odds ratios (ORs) of having metabolic syndrome increased with the selenium quartile groups (*p* for trend: <0.05), especially in women. The study demonstrated that the serum selenium levels were positively associated with metabolic syndrome following a non-linear dose–response trend. Selenium concentration was positively associated with insulin resistance in men and women, but it was associated with adiposity and lipid metabolism in women. The mechanism behind this warrants further confirmation.

Keywords: selenium; metabolic syndrome; obesity; insulin resistance; lipid

---

1. Introduction

Selenium (Se) is an antioxidative micronutrient that activates Se-containing proteins known as selenoproteins [1,2]. Among identified selenoproteins, glutathione peroxidase and selenoprotein P are more notable for their known functions of antioxidation and anti-inflammation [3,4]. Therefore, numerous investigations focused on the beneficial effects of Se exposure have tried to link it to cardiometabolic outcomes, with the emphasis mainly on type 2 diabetes (T2DM) [5–9]. Observational studies have shown a linear trend between risk of T2DM and Se exposure—both the serum Se level and dietary Se intake [10,11] but not the nail Se concentration [12]. In a meta-analysis summarizing five randomized controlled trials, a higher relative risk of T2DM in the Se-supplemented group than in the placebo group was reported [9]. When stratifying by gender, the association remained significant in men but not in women [13]. Also, the optimal range of Se exposure is narrow and may follow a non-linear, dose–response pattern [7,14].

Although studies focusing on Se and diabetes are flourishing, little is known about the association between Se and metabolic syndrome (MetS), and the conclusions remain controversial [15–18]. Although a few studies identified positive associations between the serum Se concentration and MetS only in women [15,16], or no gender differences [17], there was no significant association between serum Se concentration and MetS in the third National Health and Nutrition Examination Survey (NHANES) [18]. For obesity and dyslipidemia, general and central adiposity were negatively associated with Se levels in the NHANES [19]. Conversely, a high serum Se level was associated with increased total and non-high-density lipoprotein (HDL) cholesterol in cross-sectional studies [20,21]. However, in randomized controlled trials, Se supplementation was beneficial for decreasing total cholesterol and the total-HDL cholesterol ratio [22], or there was no significant effect between the Se-supplemented group and the placebo group [23].

Metabolic syndrome is a mixed and composite index for cardiometabolic outcomes, implying that the association between MetS and Se is complicated but deserves more detailed investigation. Therefore, we conducted this study to examine the relationship between serum Se level and MetS as well as each metabolic factor. Also, the study aimed to find a correlation between obesity, insulin resistance, and gender.

2. Materials and Methods

2.1. Study Subjects

We conducted a case control study to compare the serum Se levels between patients with and without MetS from 2007 to 2017 at the National Taiwan University Hospital. Patients who came to the outpatient department with diabetes, hypertension, hyperlipidemia, or other chronic diseases and were capable of understanding and signing the informed consent sheet were invited. A total of 1165 ambulatory males or females, aged more than 40 years, were enrolled in our study. Information about age, gender, smoking, alcohol consumption, physical activity, current medications, and previous diseases was obtained by individual interviews through questionnaires. Current smokers were defined as those smoking for more than 6 months prior to this study. Former smokers were defined as those who had not smoked for more than 12 months. Former smokers and non-smokers were grouped together as non-current smokers. Also, current alcohol drinkers were defined as those drinking more than 1 ounce of alcohol per week in the 6 months prior to this study. Former drinkers were defined as those who had quit alcohol for more than 12 months. Former drinkers and teetotalers were grouped together as non-current drinkers. Physical activity was recorded as regular exercise or not. Weight, height, systolic blood pressure (BP), and diastolic BP were measured respectively by a standard electronic scale of stadiometer and sphygmomanometer. Waist circumference (WC) was measured by a trained operator. Diabetes, hypertension, and hyperlipidemia were defined based on a self-reported history or current medication being used for those conditions. This study was approved by the Ethics Committee of National Taiwan University Hospital (201511039RINA), and written informed consent was obtained from all participants.

2.2. Definition of Metabolic Syndrome

Participants were considered to have MetS if they met three or more of the following criteria: WC $\geq$ 90 cm in men or $\geq$80 cm in women; serum triglycerides (TGs) $\geq$1.69 mmol/L; HDL cholesterol <1.03 mmol/L in men or <1.29 mmol/L in women; systolic BP $\geq$ 130 and/or diastolic BP $\geq$85 mmHg; and fasting glucose $\geq$5.56 mmol/L. Participants with medications for diabetes, hypertension, or hyperlipidemia were sorted into the group that met the criteria for fasting glucose $\geq$5.56 mmol/L, BP $\geq$ 130/85 mmHg, or serum TG $\geq$ 1.69 mmol/L, respectively.

2.3. Blood Analysis

Venous blood samples were taken after a minimum eight-hour fasting period. Serum glucose, total cholesterol, HDL cholesterol, low-density lipoprotein (LDL) cholesterol, and TG were assessed by an automatic spectrophotometric assay (HITACHI 7250, Denka Seiken Co, Niigata, Japan). Fasting insulin level was measured by a microparticle enzyme immunoassay using an AxSYM system (Abbott Laboratories, Dainabot Co, Tokyo, Japan). The homeostatic model assessment insulin resistance (HOMA-IR) was applied as an indirect measure of the degree of insulin resistance (HOMA-IR = fasting insulin × fasting plasma glucose/22.5, with glucose in mmol/L and insulin in mU/L) [24]. Serum Se was measured using inductively coupled plasma mass spectroscopy. Serum samples were diluted 1:24 with diluents of 0.1% nitric acid and 0.1% Triton X-100. The calibration standards were prepared in a blank matrix and run using the standard addition calibration type. The serum samples were analyzed in the peak-jumping mode for ^{82}Se, with the detection limit set at 0.01 µmol/L. Accuracy of the analysis was checked against Seronorm Trace Element Human Serum (batch 704121; Nycomed AS, Oslo, Norway) as reference material [6].

2.4. Statistical Analysis

Participants were divided into quartiles according to the serum Se levels. Data are presented as means (SDs) for continuous variables and numbers (percentage) for categorical variables. Multiple logistic regression analyses were performed to estimate the odds of having MetS among the quartiles of Se after adjusting for age, gender, current smoking status, current drinking status, physical activity, body mass index (BMI), and HOMA-IR. Tests for trends across serum Se quartiles were calculated by entering the quartile as an ordinal number in a regression model. Multiple linear regression models with each metabolic factor as dependent variables and serum Se as an independent variable were applied. Log transformation of the variables was performed if they were not normally distributed as assessed by the Kolmogorov–Smirnov test. The least square means were computed by general linear models adjusted for age, gender, current smoking status, current drinking status, and physical activity to compare serum Se concentration to the number of metabolic factors. Statistical analyses were performed using SPSS statistical software (V.17, SPSS, Chicago, IL, USA). A *p* value of <0.05 was considered to be statistically significant.

3. Results

The basic characteristics of the participants are shown in Table 1. The average age of the participants was 65.8 ± 10.0 years, and 64.1% were female. The mean serum Se concentration was 96.34 ± 25.90 µg/L, and the interquartile cut-off values of Se were 76.0, 94.0, and 113.7 µg/L. The serum Se levels in MetS and non-MetS groups were 102.93 ± 26.46 µg/L and 85.88 ± 21.26 µg/L, respectively. The associations of serum Se levels and prevalence of MetS by multiple logistic regression analyses are shown in Table 2. In model 1, the results showed that a higher serum Se level was correlated with a higher risk of MetS. The odds ratios (ORs) of having MetS in the second, third, and fourth Se quartile groups were 1.41 (95% CI 1.01–1.95), 2.57 (95% CI 1.83–3.59), and 5.47 (95% CI 3.75–7.96), respectively, compared with the first quartile group of serum Se level (*p* for trend: <0.001). In model 2, the results showed that a higher serum Se level was correlated with a higher risk of MetS after adjusting for age, gender, current smoking status, current drinking status, and physical activity. The ORs of having MetS in the second, third, and fourth Se quartile groups were 1.42 (95% CI 1.02–1.98), 2.39 (95% CI 1.69–3.37), and 4.96 (95% CI 3.39–7.28), respectively, compared with the first quartile (*p* for trend: <0.001). In model 3, after further adjusting for BMI, the ORs of risk for MetS in the second, third, and fourth Se quartile groups were to 1.18 (95% CI 0.80–1.73), 1.98 (95% CI 1.33–2.96), and 3.93 (95% CI 2.54–6.09), respectively, compared with the first quartile (*p* for trend: <0.001). In model 4, after further adjusting for HOMA-IR, the ORs of having MetS in the second, third, and fourth Se quartile groups decreased to 0.82 (95% CI 0.52–1.30), 1.69 (95% CI 1.03–2.79), and 1.66 (95% CI 0.88–3.12), respectively, compared with the first quartile (*p* for trend: <0.022). The interaction between Se groups and HOMA-IR was not significant (*p* = 0.057). The serum Se concentration was positively associated with WC, systolic BP, natural logarithm of TG (lnTG), fasting glucose, and HOMA-IR using multivariate linear regression analyses after adjusting for age, gender, current smoking status, current drinking status, exercise, and BMI (see Table 3).

Table 1. Characteristics of the study population by quartiles of serum selenium levels.

	Quartiles of Serum Selenium Levels			
	Q1 (*n* = 292) (≤76.0 µg/L)	Q2 (*n* = 290) (76.1–94.0 µg/L)	Q3 (*n* = 292) (94.1–113.7 µg/L)	Q4 (*n* = 291) (>113.7 µg/L)
Gender				
Female (%)	208 (71.2)	208 (71.7)	167 (57.2)	164 (56.4)
Male (%)	84 (28.8)	82 (28.3)	125 (42.8)	127 (43.6)
Age (years)	65.8 ± 10.3	65.9 ± 9.7	66.7 ± 9.6	64.9 ± 10.3
BMI (kg/m^2)	24.1 ± 3.5	24.8 ± 4.1	25.5 ± 4.3	26.5 ± 4.5
WC (cm)	82.9 ± 9.3	85.3 ± 10.6	87.6 ± 11.0	90.8 ± 11.1
Systolic BP	127.1 ± 16.8	128.0 ± 14.7	131.4 ± 15.6	159.6 ± 9.0
Diastolic BP	75.6 ± 11.0	76.2 ± 9.2	76.2 ± 10.1	68.0 ± 14.9
TCHO (mmol/L)	5.28 ± 0.95	5.05 ± 1.04	4.90 ± 1.03	4.59 ± 0.98
TGs (mmol/L)	1.51 ± 0.93	1.58 ± 1.29	1.59 ± 0.81	1.77 ± 1.23
HDL-C (mmol/L)	1.35 ± 0.32	1.33 ± 0.31	1.29 ± 0.35	1.25 ± 0.33
LDL-C (mmol/L)	3.20 ± 0.74	3.00 ± 0.80	2.94 ± 0.81	2.68 ± 0.80
Glu (mmol/L)	5.89 ± 1.47	6.22 ± 1.66	6.76 ± 2.06	7.23 ± 2.21
Insulin (U/mL)	8.30 ± 5.86	9.35 ± 7.71	10.68 ± 8.02	13.12 ± 8.71
HOMA-IR	2.28 ± 2.18	2.49 ± 2.74	3.09 ± 2.83	3.49 ± 3.07
Selenium (µg/L)	65.13 ± 7.81	85.16 ± 5.19	104.46 ± 5.59	130.66 ± 14.82
Cigarette (%)	15 (5.1)	27 (9.3)	42 (14.4)	55 (18.9)
Alcohol (%)	21 (7.2)	32 (11.0)	46 (15.8)	53 (18.2)
Exercise (%)	192 (65.8)	199 (68.6)	180 (61.6)	163 (56.0)
Diabetes (%)	76 (26.0)	114 (39.3)	183 (62.7)	247 (84.9)
Hypertension (%)	117 (40.1)	138 (47.6)	192 (65.8)	210 (75.3)
Hyperlipidemia (%)	91 (31.2)	129 (44.5)	167 (57.2)	219 (75.3)
Elevated WC (%) *	141 (48.3)	172 (59.3)	173 (59.2)	220 (75.6)
High TG (%) *	141 (48.3)	163 (56.2)	193 (66.1)	220 (75.6)
Low HDL-C (%) *	111 (38.0)	114 (39.3)	112 (38.4)	132 (45.4)
Elevated BP (%) *	134 (45.9)	139 (47.9)	183 (62.7)	186 (63.9)
IFG (%) *	136 (46.6)	170 (58.6)	213 (72.9)	256 (88.0)
Metabolic factors	2.27 ± 1.49	2.61 ± 1.48	2.99 ± 1.39	3.49 ± 1.20
MetS (%)	129 (44.2)	151 (52.1)	194 (66.4)	235 (80.8)

Abbreviations: BMI: body mass index; WC: waist circumference; BP: blood pressure; TCHO: total cholesterol; TGs: triglycerides; HDL-C: high-density lipoprotein cholesterol; LDL-C: low-density lipoprotein cholesterol; Glu: fasting glucose; HOMA-IR: homeostasis model assessment of insulin resistance; IFG: impaired fasting glucose; MetS: metabolic syndrome. * Elevated WC: WC ≥90 cm in men or ≥80 cm in women; High TG: serum TG ≥1.69 mmol/L; Low HDL-C: HDL-C <1.03 mmol/L in men or <1.29 mmol/L in women; Elevated BP: systolic BP ≥130 and/or diastolic BP ≥85 mmHg; and IFG: impaired fasting glucose ≥5.56 mmol/L. Continuous variables are presented by mean ± SD and categorical variables are presented as the percentage of participants (%).

Table 2. Odds ratios (ORs) of having MetS derived from multiple logistic regression analyses in quartiles of serum selenium levels.

	Quartile of Serum Selenium Levels				
	Q1 (*n* = 292) (≤ 76.0 µg/L)	Q2 (*n* = 290) (76.1–94.0 µg/L)	Q3 (*n* = 292) (94.1–113.7 µg/L)	Q4 (*n* = 291) (>113.7 µg/L)	*p*-value of Se Tertile
MetS, *n* (%)	129 (44.2)	151 (52.1)	194 (66.4)	235 (80.8)	
Model 1	1.00	1.41 (1.01–1.95)	2.57 (1.83–3.59)	5.47 (3.75–7.96)	<0.001
Model 2	1.00	1.42 (1.02–1.98)	2.39 (1.69–3.37)	4.96 (3.39–7.28)	<0.001
Model 3	1.00	1.18 (0.80–1.73)	1.98 (1.33–2.96)	3.93 (2.54–6.09)	<0.001
Model 4	1.00	0.82 (0.52–1.30)	1.69 (1.03–2.79)	1.66 (0.88–3.12)	0.022

Model 1: No adjustment; Model 2: adjusted for age, gender, current smoking status, current drinking status, and physical activity; Model 3 adjusted for variables in model 2, plus BMI as a confounding factor. Odds ratio of BMI (95% confidence interval [CI] 1.41–1.58, *p* < 0.001); Model 4: adjusted for variables in model 3, plus HOMA-IR as a confounding factor. Odds ratio of elevated HOMA-IR (95% CI 1.94–2.90, *p* < 0.001); HOMA-IR: homeostasis model assessment of insulin resistance; MetS: metabolic syndrome.

Table 3. Linear regression models showing standardized betas with serum selenium concentrations as independent variable for metabolic factors.

	WC		Systolic BP		Diastolic BP		lnTG		HDL-C		Fasting Glucose		HOMA-IR	
	Beta	*p*	Beta	*p*	Beta	*p*	Beta	*p*	Beta	*p*	Beta	*p*	Beta	*p*
Model 1	0.260	<0.001	0.159	<0.001	0.052	0.076	0.127	<0.001	−0.070	0.021	0.252	<0.001	0.172	<0.001
Model 2	0.284	<0.001	0.118	<0.001	0.029	0.313	0.075	0.010	−0.002	0.940	0.210	<0.001	0.135	0.001
Model 3	0.231	<0.001	0.119	<0.001	0.026	0.363	0.067	0.022	0.005	0.873	0.204	<0.001	0.132	0.001

Abbreviations: WC: waist circumference; BP: blood pressure; lnTG: natural logarithm of TG; HDL-C: high-density lipoprotein cholesterol; HOMA-IR: homeostasis model assessment of insulin resistance. Model 1: adjusted for age, gender; Model 2: adjusted for age, gender, current smoking status, current drinking status, and physical activity; Model 3: adjusted for age, gender, current smoking status, current drinking status, physical activity, and BMI.

After stratifying by gender, there was a similar higher crude OR of having MetS across the quartile groups of Se level in men (Q2: 1.89, 95% CI: 1.01–3.54; Q3: 2.32, 95% CI: 1.31–4.13; Q4: 3.63, 95% CI: 1.99–6.64, *p* for trend: <0.001) and in women (Q2: 1.26, 95% CI: 0.86–1.86; Q3: 2.57, 95%: 1.68–3.92; Q4: 7.00, 95% CI: 4.26–11.50, *p* for trend: <0.001). After adjusting for age, current smoking status, current drinking status, physical activity, and BMI, the significant trend of having a higher risk of MetS was decreased but was persistently noted more in women (Q2: 1.03, 95% CI: 0.64–1.65; Q3: 2.10, 95% CI: 1.25–3.52; Q4: 5.33, 95% CI: 2.94–9.66, *p* for trend: <0.001) than in men (Q2: 1.62, 95% CI: 0.79–3.31; Q3: 1.94, 95% CI: 0.99–3.82; Q4: 2.38, 95% CI: 1.18–4.83; *p* for trend: 0.015) (Table 4 and Figure 1). For each metabolic factor, there was a positive association with WC, systolic BP, lnTG, fasting glucose, and HOMA-IR in women, but there were only positive associations with fasting glucose and HOMA-IR in men after adjusting for age, current smoking status, current drinking status, exercise, and BMI (Table 5 and Figure 1).

Table 4. Odds ratios (ORs) of having MetS derived from multiple logistic regression analyses in quartiles of serum selenium levels, stratified by gender.

	Quartile of Serum Selenium Levels				
	Q1 (*n* = 292) (≤76.0)	Q2 (*n* = 290) (76.1–94.0)	Q3 (*n* = 292) (94.1–113.7)	Q4 (*n* = 291) (>113.7)	*p*-value of Se
Female					
MetS, *n* (%)	87/207 (40.2)	107/205 (47.8)	108/166 (65.1)	137/164 (83.5)	
Model 1	1.00	1.26 (0.86–1.86)	2.57 (1.68–3.92)	7.00 (4.26–11.50)	<0.001
Model 2	1.00	1.20 (0.81–1.80)	2.38 (1.55–3.66)	6.29 (3.78–10.45)	<0.001
Model 3	1.00	1.03 (0.64–1.65)	2.10 (1.25–3.52)	5.33 (2.94–9.66)	<0.001
Male					
MetS, *n* (%)	42/84 (50)	53/81 (65.4)	86/123 (69.9)	98/125 (78.4)	
Model 1	1.00	1.89 (1.01–3.54)	2.32 (1.31–4.13)	3.63 (1.99–6.64)	<0.001
Model 2	1.00	2.14 (1.10–4.15)	2.59 (1.40–4.79)	3.08 (1.63–5.83)	0.001
Model 3	1.00	1.62 (0.79–3.31)	1.94 (0.99–3.82)	2.38 (1.18–4.83)	0.015

Model 1: adjusted for age; Model 2: adjusted for age, current smoking status, current drinking status, and physical activity; Model 3 adjusted for variables in model 2, plus BMI as a confounding factor. Odds ratio of BMI (95% confidence interval [CI] 1.44–1.68, *p* < 0.001 for male; 95% CI 1.26–1.51, *p* < 0.001 for female). MetS: metabolic syndrome.

The least square means (± SDs) of serum Se concentration in relation to the number of metabolic factors are shown in Figure 2A. In the linear multiple regression models, after adjusting for age, gender, current smoking status, current drinking status, and physical activity, the serum Se concentration increased with the escalation of the number of metabolic factors (test for trend: *p* < 0.001). After stratifying by gender, the serum Se concentration increased as the number of metabolic factors increased both in female and male patients after adjustment (test for trend: *p* < 0.001) (Figure 2B,C).

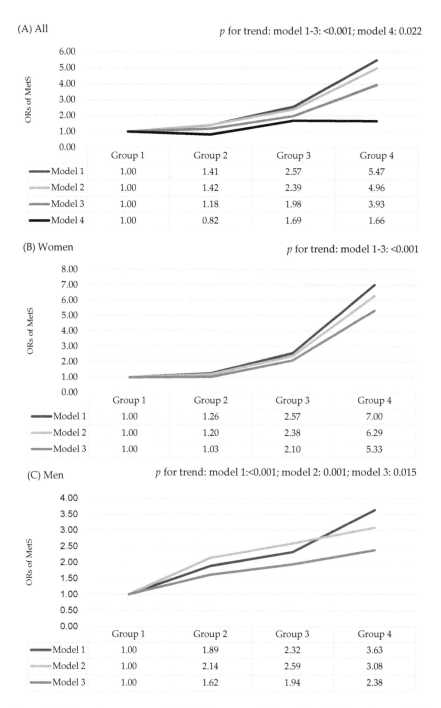

Figure 1. Nonlinear dose–response relationship between selenium and metabolic syndrome. (**A**) All subjects; (**B**) Female subjects; (**C**) Male subjects.

Table 5. Linear regression models showing standardized betas with serum selenium concentrations as independent variable for metabolic factors, stratified by gender.

	WC		Systolic BP		Diastolic BP		lnTG		HDL-C		Fasting Glucose		HOMA-IR	
	Beta	*p*	Beta	*p*	Beta	*p*	Beta	*p*	Beta	*p*	Beta	*p*	Beta	*p*
Female														
Model 1	0.266	<0.001	0.197	<0.001	0.077	0.035	0.184	<0.001	−0.067	<0.001	0.271	<0.001	0.192	<0.001
Model 2	0.234	<0.001	0.184	<0.001	0.095	0.010	0.164	<0.001	−0.049	0.186	0.219	<0.001	0.166	<0.001
Model 3	0.056	0.002	0.139	<0.001	0.069	0.064	0.108	0.003	0.012	0.738	0.162	<0.001	0.083	0.035
Male														
Model 1	0.228	<0.001	0.072	0.141	0.022	0.450	0.039	0.430	−0.078	0.111	0.230	<0.001	0.129	0.048
Model 2	0.168	0.001	0.076	0.142	−0.028	0.580	−0.027	0.589	−0.072	0.155	0.171	0.009
Model 3	0.048	0.056	0.051	0.321	−0.048	0.349	−0.052	0.295	−0.010	0.843	0.211	<0.001	0.158	0.008

Abbreviations: WC: waist circumference; BP: blood pressure; HDL-C: high-density lipoprotein cholesterol; HOMA-IR: homeostasis model assessment of insulin resistance. Model 1: adjusted for age; Model 2: adjusted for age, current smoking status, current drinking status, and physical activity; Model 3: adjusted for age, current smoking status, current drinking status, physical activity, and BMI.

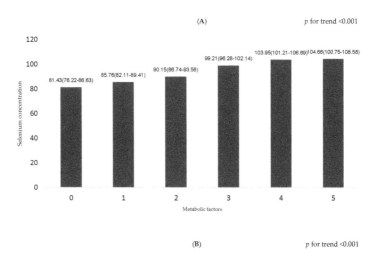

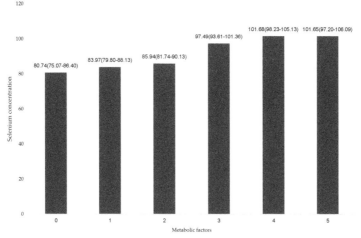

Figure 2. *Cont.*

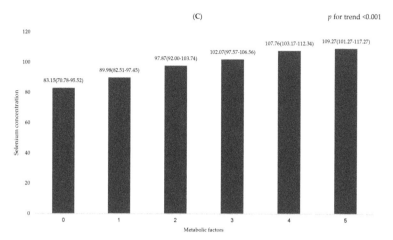

Figure 2. Comparison of serum selenium concentration in relation to number of metabolic factors. (**A**) All subjects; (**B**) Female subjects; (**C**) Male subjects.

4. Discussion

The results of the present study showed a positive association between serum Se level and the risk of MetS. Also, the serum Se concentration was positively associated with WC, systolic BP, lnTG, fasting glucose, HOMA-IR, and the number of metabolic factors (test for trend: $p < 0.001$), following a dose–response relationship. Further, there was a 3.93-fold risk of MetS in the highest Se quartile compared with the lowest quartile after adjusting for demographic confounders and BMI (5.33-fold in women and 2.38-fold in men). Although further adjustment for HOMA-IR diminished most of the magnitude of the association between Se and MetS, there was a non-linear, dose–response trend whereby the odds of having MetS with the escalation of Se level (p for trend 0.022). These findings support a positive association between serum Se gradients and MetS independent of obesity and insulin resistance. The persistence of a direct relationship between Se exposure and risk of MetS after adjusting for BMI and HOMA-IR also implied that as-yet-unidentified confounding variables affected this association. Stratifying by gender, Se level was positively associated with insulin resistance (fasting glucose and HOMA-IR) in men and women, but with adiposity and lipid metabolism (WC, SBP, and lnTG) in women only, implying an effect modification by dimorphic gender.

The overall findings of this study are in agreement with the majority of previous observational studies, which reported positive associations between serum Se level and MetS [15–17]. In a Chinese case-control study, a higher level of plasma Se was associated with an increased risk of MetS both in men and women [15]. Similarly, the IMMIDIET (The dietary habit profile in European communities with different risk of myocardial infarction: the impact of migration as a model of gene-environment interaction) project and an observational study in Lebanon showed a positive association between serum Se and MetS, but only in women [16,17]. Conversely, there was no significant association between serum Se and MetS in the third NHANES [18]. In animal models, knockout mice under adequate Se diets developed MetS pattern including hyperinsulinemia, increased body weight, dyslipidemia, and glucose intolerance [25]. The potential mechanism to link Se to insulin resistance and obesity may be partly mediated by glutathione peroxidase and selenoprotein P, due to their notable anti-inflammation functions [7]. Associated with the attenuation of antioxidative actions, Se-supplemented rats were found to develop insulin resistance [26]. Also, the overexpression of glutathione peroxidase induced the development of insulin resistance and obesity in mice [27]. Furthermore, there was an association between gene polymorphism of selenoprotein P and fasting insulin in a human study, supporting the role of selenoprotein P in glucose metabolism [28]. In terms of gender differences, glutathione peroxidase overexpression with hyperinsulinemia was only observed in male mice [29], whereas the expression of glutathione peroxidase in liver was observed more in female-derived cells compared to male-derived cells [30] Moreover, elevated selenoprotein P and insulin resistance were only observed in female mice [31]. In terms of human gene investigations, there was an elevated expression of glutathione peroxidase and selenoprotein P genes in women in relation to obesity in the England SELGEN study [32], while glutathione peroxidase polymorphisms were related to an increased incidence of MetS in men in a Japanese adult cohort [33]. In a Finnish cohort, variation in the selenoprotein S gene locus was associated with coronary heart disease and ischemic stroke in women [34]. There were gender differences in the amount of Se necessary to reach optimal Se expression.

A meta-analysis that pooled five observational studies of 13,460 subjects found that there was a positive, non-linear, dose–response association between serum Se levels and T2DM [14]. Also, there was a higher relative risk of T2DM in the Se-supplemented group than the placebo group in a meta-analysis summarizing five randomized controlled trials [9]. In animal studies, both overexpression and deficiency of selenoproteins can promote the development of T2DM, following a non-linear correlation [35]. For adiposity, serum Se was inversely associated with BMI in both men and women, whereas it was associated with the percentage of body fat only in women in the third NHANES [19]. However, there were consistent and positive associations between the Se concentration and total cholesterol [21,36], TG [37], and non-HDL cholesterol [21]. These outcomes were also identified in the UK PRECISE (Prevention of Cancer by Intervention with Selenium) study [22] but not in a study of an elderly Danish population's [23] risk of elevated lipid profiles in the Se-supplementation group. Little is known about the relationship between Se and hypertension. Although some studies have showed positive associations between Se and systolic and diastolic BP [21,34], there was no association shown between Se and hypertension in a systemic review [38]. Generally speaking, previous observational and randomized controlled studies have elucidated a positive trend between Se and MetS, but the differences between genders are still debated. This might be related to unequal organ distribution [30], different optimal levels for Se expression [7,29,31], and polymorphisms in different genders [32–34]. In our study, we further confirmed the dose–response association between Se and MetS and found that dimorphic genders differed in response to insulin resistance, adiposity, and lipid profiles in relation to the Se level.

There are some limitations to our study. First, we were not able to establish the causal relationship between serum Se concentration and MetS because of the cross-sectional design. Although we collected and adjusted for probable confounders in our study, there could be unmeasured and undefined factors with possible residual effects. For example, there were potential influences of the duration of cardiometabolic diseases on lowering the serum Se level over time, but we did not measure the time elapsed during the development of metabolic factors among individuals with or without MetS. Moreover, the serum Se level could be altered by dietary sources of Se, including soybeans, bamboo shoots, broccoli, mushrooms, cereals, Brazil nuts, and milk powder [39]. Because we did not record the daily micronutrient supplementation and personal eating habits, there might be bias independent of MetS. We checked the total serum Se to represent the serum selenoprotein concentration and activity, but we did not determine the proportion of other forms of Se and their activities. Furthermore, we used HOMA-IR as an indirect approach to estimate the degree of insulin resistance instead of accurate dynamic techniques, such as using a euglycemic clamp. Nonetheless, this is the first human study to comprehensively demonstrate the dose–response relationship between Se and MetS and metabolic factors with a large sample size. Gender stratification analyses clearly highlighted the gender differences in insulin resistance, adiposity, and lipid metabolism. However, the underlying mechanisms need further investigation.

Author Contributions: K.-C.H. conceptualized and designed the study, secured funding for the study. C.-W.L. and K.-C.Y. performed the formal analyses; C.-W.L. and H.-H.C. prepared the original draft; K.-C.Y., C.-H.C., C.-A.Y., and K.-C.H. contributed medical expertise and reviewed the manuscript. All authors read and approved the final version of the manuscript.

Funding: This work was supported in part by the National Health Research Institutes, Taiwan (Grant numbers: PH-107-PP-23).

Acknowledgments: The authors would like to thank Ms. Y.H. Lin and Ms. H.F. Hu for help in questionnaire collection, data archiving, and administrative support.

Conflicts of Interest: The authors declare no conflict of interest. The funders had no role in the design of the study; in the collection, analyses or interpretation of data; in the writing of the manuscript or in the decision to publish the results.

References

1. Rayman, M.P. Selenium and human health. *Lancet* **2012**, *379*, 1256–1268. [CrossRef]
2. Reeves, M.A.; Hoffmann, P.R. The human selenoproteome: Recent insights into functions and regulation. *Cell. Mol. Life Sci.* **2009**, *66*, 2457–2478. [CrossRef] [PubMed]
3. Burk, R.F.; Hill, K.E. Selenoprotein P-expression, functions, and roles in mammals. *Biochim. Biophys. Acta* **2009**, *1790*, 1441–1447. [CrossRef] [PubMed]
4. Matsuda, M.; Shimomura, I. Increased oxidative stress in obesity: Implications for metabolic syndrome, diabetes, hypertension, dyslipidemia, atherosclerosis, and cancer. *Obes. Res. Clin. Pract.* **2013**, *7*, e330–e341. [CrossRef] [PubMed]
5. Stranges, S.; Galletti, F.; Farinaro, E.; D'Elia, L.; Russo, O.; Iacone, R.; Capasso, C.; Carginale, V.; De Luca, V.; Della Valle, E.; et al. Associations of selenium status with cardiometabolic risk factors: An 8-year follow-up analysis of the Olivetti Heart study. *Atherosclerosis* **2011**, *217*, 274–278. [CrossRef] [PubMed]
6. Lu, C.W.; Chang, H.H.; Yang, K.C.; Kuo, C.S.; Lee, L.T.; Huang, K.C. High serum selenium levels are associated with increased risk for diabetes mellitus independent of central obesity and insulin resistance. *BMJ Open Diabetes Res. Care* **2016**, *4*, e000253. [CrossRef] [PubMed]
7. Ogawa-Wong, A.N.; Berry, M.J.; Seale, L.A. Selenium and Metabolic Disorders: An Emphasis on Type 2 Diabetes Risk. *Nutrients* **2016**, *8*, 80. [CrossRef] [PubMed]
8. Kohler, L.N.; Florea, A.; Kelley, C.P.; Chow, S.; Hsu, P.; Batai, K.; Saboda, K.; Lance, P.; Jacobs, E.T. Higher Plasma Selenium Concentrations Are Associated with Increased Odds of Prevalent Type 2 Diabetes. *J. Nutr.* **2018**, *148*, 1333–1340. [CrossRef] [PubMed]
9. Vinceti, M.; Filippini, T.; Rothman, K.J. Selenium exposure and the risk of type 2 diabetes: A systematic review and meta-analysis. *Eur. J. Epidemiol.* **2018**, *33*, 789–810. [CrossRef] [PubMed]
10. Christensen, K.; Werner, M.; Malecki, K. Serum selenium and lipid levels: Associations observed in the National Health and Nutrition Examination Survey (NHANES) 2011–2012. *Environ. Res.* **2015**, *140*, 76–84. [CrossRef] [PubMed]
11. Stranges, S.; Sieri, S.; Vinceti, M.; Grioni, S.; Guallar, E.; Laclaustra, M.; Muti, P.; Berrino, F.; Krogh, V. A prospective study of dietary selenium intake and risk of type 2 diabetes. *BMC Public Health* **2010**, *10*, 564. [CrossRef] [PubMed]
12. Park, K.; Rimm, E.B.; Siscovick, D.S.; Spiegelman, D.; Manson, J.E.; Morris, J.S.; Hu, F.B.; Mozaffarian, D. Toenail selenium and incidence of type 2 diabetes in U.S. men and women. *Diabetes Care* **2012**, *35*, 1544–1551. [CrossRef] [PubMed]
13. Stranges, S.; Marshall, J.R.; Natarajan, R.; Donahue, R.P.; Trevisan, M.; Combs, G.F.; Cappuccio, F.P.; Ceriello, A.; Reid, M.E. Effects of long-term selenium supplementation on the incidence of type 2 diabetes: A randomized trial. *Ann. Intern. Med.* **2007**, *147*, 217–223. [CrossRef] [PubMed]
14. Wang, X.L.; Yang, T.B.; Wei, J.; Lei, G.H.; Zeng, C. Association between serum selenium level and type 2 diabetes mellitus: A non-linear dose-response meta-analysis of observational studies. *Nutr. J.* **2016**, *15*, 48. [CrossRef] [PubMed]
15. Yuan, Z.; Xu, X.; Ye, H.; Jin, L.; Zhang, X.; Zhu, Y. High levels of plasma selenium are associated with metabolic syndrome and elevated fasting plasma glucose in a Chinese population: A case-control study. *J. Trace. Elem. Med. Biol.* **2015**, *32*, 189–194. [CrossRef] [PubMed]

16. Arnaud, J.; de Lorgeril, M.; Akbaraly, T.; Salen, P.; Arnout, J.; Cappuccio, F.P.; van Dongen, M.C.; Donati, M.B.; Krogh, V.; Siani, A.; et al. Gender differences in copper, zinc and selenium status in diabetic-free metabolic syndrome European population—The IMMIDIET study. *Nutr. Metab. Cardiovasc. Dis.* **2012**, *22*, 517–524. [CrossRef] [PubMed]

17. Obeid, O.; Elfakhani, M.; Hlais, S.; Iskandar, M.; Batal, M.; Mouneimne, Y.; Adra, N.; Hwalla, N. Plasma copper, zinc, and selenium levels and correlates with metabolic syndrome components of lebanese adults. *Biol. Trace. Elem. Res.* **2008**, *123*, 58–65. [CrossRef] [PubMed]

18. Ford, E.S.; Mokdad, A.H.; Giles, W.H.; Brown, D.W. The metabolic syndrome and antioxidant concentrations: Findings from the Third National Health and Nutrition Examination Survey. *Diabetes* **2003**, *52*, 2346–2352. [CrossRef] [PubMed]

19. Zhong, Q.; Lin, R.; Nong, Q. Adiposity and Serum Selenium in U.S. Adults. *Nutrients* **2018**, *10*, 727. [CrossRef] [PubMed]

20. Gonzalez-Estecha, M.; Palazon-Bru, I.; Bodas-Pinedo, A.; Trasobares, E.; Palazon-Bru, A.; Fuentes, M.; Cuadrado-Cenzual, M.A.; Calvo-Manuel, E. Relationship between serum selenium, sociodemographic variables, other trace elements and lipid profile in an adult Spanish population. *J. Trace. Elem. Med. Biol.* **2017**, *43*, 93–105. [CrossRef] [PubMed]

21. Stranges, S.; Laclaustra, M.; Ji, C.; Cappuccio, F.P.; Navas-Acien, A.; Ordovas, J.M.; Rayman, M.; Guallar, E. Higher selenium status is associated with adverse blood lipid profile in British adults. *J. Nutr.* **2010**, *140*, 81–87. [CrossRef] [PubMed]

22. Rayman, M.P.; Stranges, S.; Griffin, B.A.; Pastor-Barriuso, R.; Guallar, E. Effect of supplementation with high-selenium yeast on plasma lipids: A randomized trial. *Ann. Intern. Med.* **2011**, *154*, 656–665. [CrossRef] [PubMed]

23. Cold, F.; Winther, K.H.; Pastor-Barriuso, R.; Rayman, M.P.; Guallar, E.; Nybo, M.; Griffin, B.A.; Stranges, S.; Cold, S. Randomised controlled trial of the effect of long-term selenium supplementation on plasma cholesterol in an elderly Danish population. *Br. J. Nutr.* **2015**, *114*, 1807–1818. [CrossRef] [PubMed]

24. Matthews, D.R.; Hosker, J.P.; Rudenski, A.S.; Naylor, B.A.; Treacher, D.F.; Turner, R.C. Homeostasis model assessment: Insulin resistance and beta-cell function from fasting plasma glucose and insulin concentrations in man. *Diabetologia* **1985**, *28*, 412–419. [CrossRef] [PubMed]

25. Seale, L.A.; Hashimoto, A.C.; Kurokawa, S.; Gilman, C.L.; Seyedali, A.; Bellinger, F.P.; Raman, A.V.; Berry, M.J. Disruption of the selenocysteine lyase-mediated selenium recycling pathway leads to metabolic syndrome in mice. *Mol. Cell. Biol.* **2012**, *32*, 4141–4154. [CrossRef] [PubMed]

26. Wang, X.; Zhang, W.; Chen, H.; Liao, N.; Wang, Z.; Zhang, X.; Hai, C. High selenium impairs hepatic insulin sensitivity through opposite regulation of ROS. *Toxicol. Lett.* **2014**, *224*, 16–23. [CrossRef] [PubMed]

27. McClung, J.P.; Roneker, C.A.; Mu, W.; Lisk, D.J.; Langlais, P.; Liu, F.; Lei, X.G. Development of insulin resistance and obesity in mice overexpressing cellular glutathione peroxidase. *Proc. Natl. Acad. Sci. USA* **2004**, *101*, 8852–8857. [CrossRef] [PubMed]

28. Hellwege, J.N.; Palmer, N.D.; Ziegler, J.T.; Langefeld, C.D.; Lorenzo, C.; Norris, J.M.; Takamura, T.; Bowden, D.W. Genetic variants in selenoprotein P plasma 1 gene (SEPP1) are associated with fasting insulin and first phase insulin response in Hispanics. *Gene* **2014**, *534*, 33–39. [CrossRef] [PubMed]

29. Wang, X.D.; Vatamaniuk, M.Z.; Wang, S.K.; Roneker, C.A.; Simmons, R.A.; Lei, X.G. Molecular mechanisms for hyperinsulinaemia induced by overproduction of selenium-dependent glutathione peroxidase-1 in mice. *Diabetologia* **2008**, *51*, 1515–1524. [CrossRef] [PubMed]

30. Schomburg, L.; Schweizer, U. Hierarchical regulation of selenoprotein expression and sex-specific effects of selenium. *Biochim. Biophys. Acta* **2009**, *1790*, 1453–1462. [CrossRef] [PubMed]

31. Misu, H.; Takamura, T.; Takayama, H.; Hayashi, H.; Matsuzawa-Nagata, N.; Kurita, S.; Ishikura, K.; Ando, H.; Takeshita, Y.; Ota, T.; et al. A liver-derived secretory protein, selenoprotein P, causes insulin resistance. *Cell. Metab.* **2010**, *12*, 483–495. [CrossRef] [PubMed]

32. Meplan, C.; Crosley, L.K.; Nicol, F.; Beckett, G.J.; Howie, A.F.; Hill, K.E.; Horgan, G.; Mathers, J.C.; Arthur, J.R.; Hesketh, J.E. Genetic polymorphisms in the human selenoprotein P gene determine the response of selenoprotein markers to selenium supplementation in a gender-specific manner (the SELGEN study). *FASEB J.* **2007**, *21*, 3063–3074. [CrossRef] [PubMed]

33. Kuzuya, M.; Ando, F.; Iguchi, A.; Shimokata, H. Glutathione peroxidase 1 Pro198Leu variant contributes to the metabolic syndrome in men in a large Japanese cohort. *Am. J. Clin. Nutr.* **2008**, *87*, 1939–1944. [CrossRef] [PubMed]

34. Alanne, M.; Kristiansson, K.; Auro, K.; Silander, K.; Kuulasmaa, K.; Peltonen, L.; Salomaa, V.; Perola, M. Variation in the selenoprotein S gene locus is associated with coronary heart disease and ischemic stroke in two independent Finnish cohorts. *Hum. Genet.* **2007**, *122*, 355–365. [CrossRef] [PubMed]

35. Zhou, J.; Huang, K.; Lei, X.G. Selenium and diabetes—Evidence from animal studies. *Free Radic. Biol. Med.* **2013**, *65*, 1548–1556. [CrossRef] [PubMed]

36. Laclaustra, M.; Stranges, S.; Navas-Acien, A.; Ordovas, J.M.; Guallar, E. Serum selenium and serum lipids in US adults: National Health and Nutrition Examination Survey (NHANES) 2003–2004. *Atherosclerosis* **2010**, *210*, 643–648. [CrossRef] [PubMed]

37. Berthold, H.K.; Michalke, B.; Krone, W.; Guallar, E.; Gouni-Berthold, I. Influence of serum selenium concentrations on hypertension: The Lipid Analytic Cologne cross-sectional study. *J. Hypertens.* **2012**, *30*, 1328–1335. [CrossRef] [PubMed]

38. Kuruppu, D.; Hendrie, H.C.; Yang, L.; Gao, S. Selenium levels and hypertension: A systematic review of the literature. *Public Health Nutr.* **2014**, *17*, 1342–1352. [CrossRef] [PubMed]

39. Rayman, M.P.; Infante, H.G.; Sargent, M. Food-chain selenium and human health: Spotlight on speciation. *Br. J. Nutr.* **2008**, *100*, 238–253. [CrossRef] [PubMed]

nutrients

Article

Association of Selenoprotein and Selenium Pathway Genotypes with Risk of Colorectal Cancer and Interaction with Selenium Status

Veronika Fedirko [1], Mazda Jenab [2], Catherine Méplan [3], Jeb S. Jones [1], Wanzhe Zhu [1], Lutz Schomburg [4], Afshan Siddiq [5], Sandra Hybsier [4], Kim Overvad [6], Anne Tjønneland [7], Hanane Omichessan [8,9], Vittorio Perduca [8,9,10], Marie-Christine Boutron-Ruault [8,9], Tilman Kühn [11], Verena Katzke [11], Krasimira Aleksandrova [12], Antonia Trichopoulou [13], Anna Karakatsani [13,14], Anastasia Kotanidou [13,15], Rosario Tumino [16], Salvatore Panico [17], Giovanna Masala [18], Claudia Agnoli [19], Alessio Naccarati [20], Bas Bueno-de-Mesquita [5,21,22,23], Roel C.H. Vermeulen [24], Elisabete Weiderpass [25,26,27,28], Guri Skeie [28], Therese Haugdahl Nøst [28], Leila Lujan-Barroso [29], J. Ramón Quirós [30], José María Huerta [31,32], Miguel Rodríguez-Barranco [32,33], Aurelio Barricarte [32,34,35], Björn Gylling [36], Sophia Harlid [37], Kathryn E. Bradbury [38], Nick Wareham [39], Kay-Tee Khaw [40], Marc Gunter [2], Neil Murphy [2], Heinz Freisling [2], Kostas Tsilidis [5,41], Dagfinn Aune [5,42,43], Elio Riboli [5], John E. Hesketh [3] and David J. Hughes [44,*]

1   Department of Epidemiology, Rollins School of Public Health & Winship Cancer Institute, Emory University, Atlanta, GA 30322, USA; veronika.fedirko@emory.edu (V.F.); jeb.jones@emory.edu (J.S.J.); WZHU4@emory.edu (W.Z.)
2   Section of Nutrition and Metabolism, International Agency for Research on Cancer, 69372 Lyon, France; jenabm@iarc.fr (M.J.); GunterM@iarc.fr (M.G.); MurphyN@iarc.fr (N.M.); FreislingH@iarc.fr (H.F.)
3   School of Biomedical Sciences, Newcastle University, Newcastle upon Tyne NE1 7RU, UK; Catherine.Meplan@newcastle.ac.uk (C.M.); j.hesketh@rgu.ac.uk (J.E.H.)
4   Institute for Experimental Endocrinology, University Medical School, D-13353 Berlin, Germany; lutz.schomburg@charite.de (L.S.); sandra.hybsier@charite.de (S.H.)
5   Department of Epidemiology and Biostatistics, The School of Public Health, Imperial College London, London W2 1PG, UK; afshan.siddiq@genomicsengland.co.uk (A.S.); bas.bueno.de.mesquita@rivm.nl (B.B.-d.-M.); ktsilidis@gmail.com (K.T.); d.aune@imperial.ac.uk (D.A.); e.riboli@imperial.ac.uk (E.R.)
6   Department of Public Health, Section for Epidemiology, Aarhus University, 8000 Aarhus, Denmark; ko@dce.au.dk
7   Diet, Genes and Environment Unit, Danish Cancer Society Research Center, DK 2100 Copenhagen, Denmark; annet@cancer.dk
8   Faculty of Medicine, CESP, University of Paris-Sud, Faculty of Medicine UVSQ, INSERM, University of Paris-Saclay, 94805 Villejuif, France; HANANE.OMICHESSAN@gustaveroussy.fr (H.O.); vittorio.perduca@gmail.com (V.P.); boutron@igr.fr (M.-C.B.-R.)
9   Centre for Research in Epidemiology and Population Health (CESP), F-94805 Gustave Roussy, Villejuif, France
10 Laboratory of Applied Mathematics, MAP5 (UMR CNRS 8145), University of Paris Descartes, 75270 Paris, France
11 Division of Cancer Epidemiology, German Cancer Research Centre (DKFZ), 69120 Heidelberg, Germany; t.kuehn@dkfz-Heidelberg.de (T.K.); v.katzke@dkfz-Heidelberg.de (V.K.)
12 Department of Epidemiology, German Institute of Human Nutrition Potsdam-Rehbrücke, 14558 Nuthetal, Germany; Krasimira.Aleksandrova@dife.de
13 Hellenic Health Foundation, 115 27 Athens, Greece; atrichopoulou@hhf-greece.gr (A.T.); a.karakatsani@hhf-greece.gr (A.K.); a.kotanidou@hhf-greece.gr (A.K.)
14 2nd Pulmonary Medicine Department, School of Medicine, National and Kapodistrian University of Athens, "ATTIKON" University Hospital, 106 79 Haidari, Greece
15 1st Department of Critical Care Medicine and Pulmonary Services, University of Athens Medical School, Evangelismos Hospital, 106 76 Athens, Greece

16   Cancer Registry and Histopathology Department, Civic M.P. Arezzo Hospital, 97100 Ragusa, Italy;
      rtumino@tin.it

17   Department of Clinical Medicine and Surgery, Federico II University, 80138 Naples, Italy; spanico@unina.it

18   Cancer Risk Factors and Life-Style Epidemiology Unit, Cancer Research and Prevention Institute—ISPO,
      50141 Florence, Italy; g.masala@ispo.toscana.it

19   Epidemiology and Prevention Unit, IRCCS Foundation National Cancer Institute, 20133 Milan, Italy;
      claudia.agnoli@istitutotumori.mi.it

20   Molecular and Genetic Epidemiology Unit, Italian Institute for Genomic Medicine (IIGM) Torino,
      10126 Torino, Italy; alessio.naccarati@hugef-torino.org

21   Department for Determinants of Chronic Diseases (DCD), National Institute for Public Health and the
      Environment (RIVM), 3720 Bilthoven, The Netherlands

22   Department of Gastroenterology and Hepatology, University Medical Centre, 3584 CX Utrecht,
      The Netherlands

23   Department of Social and Preventive Medicine, Faculty of Medicine, University of Malaya,
      Kuala Lumpur 50603, Malaysia

24   Institute of Risk Assessment Sciences, Utrecht University, 3512 JE Utrecht, The Netherlands;
      R.C.H.Vermeulen@uu.nl

25   Department of Research, Cancer Registry of Norway, Institute of Population-Based Cancer Research,
      N-0304 Oslo, Norway; WeiderpassE@iarc.fr

26   Department of Medical Epidemiology and Biostatistics, Karolinska Institute, SE-171 77 Stockholm, Sweden

27   Genetic Epidemiology Group, Folkhälsan Research Center, and Faculty of Medicine, Helsinki University,
      00014 Helsinki, Finland

28   Department of Community Medicine, University of Tromsø, The Arctic University of Norway, 9019 Tromsø,
      Norway; Guri.Skeie@ism.uit.no (G.S.); therese.h.nost@uit.no (T.H.N.)

29   Unit of Nutrition and Cancer, Catalan Institute of Oncology (ICO-IDIBELL), L'Hospitalet de Llobregat,
      08908 Barcelona, Spain; llujan@iconcologia.net

30   EPIC Asturias, Public Health Directorate, 33006 Oviedo, Asturias, Spain;
      joseramon.quirosgarcia@asturias.org

31   Department of Epidemiology, Murcia Regional Health Council, IMIB-Arrixaca, 30008 Murcia, Spain;
      jmhuerta.carm@gmail.com

32   CIBER Epidemiology and Public Health (CIBERESP), 28029 Madrid, Spain;
      miguel.rodriguez.barranco.easp@juntadeandalucia.es (M.R.-B.);
      aurelio.barricarte.gurrea@cfnavarra.es (A.B.)

33   Andalucia School of Public Health, Institute for Biosanitary Research, University Hospital of Granada,
      University of Granada, 18011 Granada, Spain

34   Epidemiology, Prevention and Promotion Health Service, Navarra Public Health Institute,
      31003 Pamplona, Spain

35   Navarra Institute for Health Research (IdiSNA), 31008 Pamplona, Spain

36   Department of Medical Biosciences, Pathology, Umea University, 901 87 Umea, Sweden;
      bjorn.gylling@umu.se

37   Department of Radiation Sciences, Oncology, Umea University, 901 87 Umea, Sweden; sophia.harlid@umu.se

38   Cancer Epidemiology Unit, Nuffield Department of Population Health, University of Oxford,
      Oxford OX3 7LF, UK; kathryn.bradbury@ceu.ox.ac.uk

39   MRC Epidemiology Unit, University of Cambridge, CB2 0QQ Cambridge, UK;
      nick.wareham@mrc-epid.cam.ac.uk

40   School of Clinical Medicine, University of Cambridge, Clinical Gerontology Unit, Addenbrooke's Hospital,
      Cambridge CB2 0QQ, UK; kk101@medschl.cam.ac.uk

41   Department of Hygiene and Epidemiology, University of Ioannina School of Medicine,
      45110 Ioannina, Greece

42   Department of Nutrition, Bjørknes University College, 0456 Oslo, Norway

43   Department of Endocrinology, Morbid Obesity and Preventive Medicine, Oslo University Hospital,
      0372 Oslo, Norway

44   Cancer Biology and Therapeutics Group, UCD Conway Institute, School of Biomolecular and Biomedical
      Science, University College Dublin, D04 V1W8 Dublin, Ireland

*   Correspondence: david.hughes@ucd.ie; Tel.: +353-1-716-6988

Received: 27 March 2019; Accepted: 22 April 2019; Published: 25 April 2019

Abstract: Selenoprotein genetic variations and suboptimal selenium (Se) levels may contribute to the risk of colorectal cancer (CRC) development. We examined the association between CRC risk and genotype for single nucleotide polymorphisms (SNPs) in selenoprotein and Se metabolic pathway genes. *Illumina Goldengate* assays were designed and resulted in the genotyping of 1040 variants in 154 genes from 1420 cases and 1421 controls within the European Prospective Investigation into Cancer and Nutrition (EPIC) study. Multivariable logistic regression revealed an association of 144 individual SNPs from 63 Se pathway genes with CRC risk. However, regarding the selenoprotein genes, only *TXNRD1* rs11111979 retained borderline statistical significance after adjustment for correlated tests (P_{ACT} = 0.10; P_{ACT} significance threshold was $P < 0.1$). SNPs in Wingless/Integrated (Wnt) and Transforming growth factor (TGF) beta-signaling genes (*FRZB, SMAD3, SMAD7*) from pathways affected by Se intake were also associated with CRC risk after multiple testing adjustments. Interactions with Se status (using existing serum Se and Selenoprotein P data) were tested at the SNP, gene, and pathway levels. Pathway analyses using the modified Adaptive Rank Truncated Product method suggested that genes and gene x Se status interactions in antioxidant, apoptosis, and TGF-beta signaling pathways may be associated with CRC risk. This study suggests that SNPs in the Se pathway alone or in combination with suboptimal Se status may contribute to CRC development.

Keywords: selenium; selenium status; selenoprotein gene variation; selenium pathway; colorectal neoplasms; selenoprotein P; prospective cohort; colorectal cancer risk; genetic epidemiology; biomarkers

---

1. Introduction

In Europe, colorectal cancer (CRC) is the cancer type with both the second highest incidence and mortality rate [1]. Substantial CRC risk may derive from dietary factors, genetic variants, and their interactions [2,3].

Experimental and observational evidence suggests that suboptimal dietary intakes of the micronutrient selenium (Se) contribute to greater risk for the development of cancers at several anatomical sites, including the colorectum [4–6]. In humans, Se exerts its potential anti-carcinogenic properties through incorporation into 25 selenoproteins by the amino acid selenocysteine [7,8]. Several selenoproteins protect cells from damaging oxidative radicals including the glutathione peroxidases (notably GPX1 and GPX4), components of the thioredoxin reductase system (TXNRD1-3) and selenoprotein P (SELENOP; please note the modified selenoprotein nomenclature [9]) which is also critical for Se transport [8,10,11].

The major mechanism through which Se is thought to influence the risk of CRC development is variation in gene expression and biosynthesis of protective selenoproteins [12,13]. In rodent models, adequate Se intake and selenoprotein expression have been shown to prevent colon cancer while selenoprotein dysregulation may increase colon cancer risk [14–16]. Data from nutritional intervention trials and epidemiological studies suggest implications for Se intake regarding CRC risk could potentially be more important in individuals with particular selenoprotein genotypes and/or in populations with low Se status, such as in Western Europe where the present study was conducted [4,5,17,18]. Risk modification by sex has also been observed for CRC risk associations with selenoprotein genotypes [19,20] and Se status [5,17].

Genetic variations in approximately half of the 25-known human selenoprotein genes have been associated with susceptibility to CRC and/or colorectal adenoma (CRA) risk in at least seven populations from Asia, North America and Europe; in addition some of these variants have been shown to impact survival outcomes (reviewed in [4,21]). Although some of these studies have been performed in suboptimal Se intake areas, large studies have more generally been conducted in Se-replete environments in North America and these have reported evidence both for [22,23] and against [17] an association of selenoprotein genes with CRC risk. However, only a limited number of single nucleotide polymorphisms (SNPs) in selected selenoprotein genes have been analyzed, while in several of these reports the Se status of the analyzed cohort was not assessed.

To our knowledge, it is unknown which selenoproteins are critical in maintaining colonic health and no study has comprehensively evaluated variation in all selenoprotein genes for association with CRC risk. Moreover, interactions of selenoprotein genetic variations according to robust Se status biomarkers have not been explored. As both genetic factors and dietary Se intake can influence the pattern of selenoprotein expression and biosynthesis, we hypothesized that variation in selenoprotein genes, and in related signaling pathway genes influenced by Se intake (together comprising the 'Se pathway'), affect CRC development risk, while Se status may modify this risk.

In this study, we have examined for the first time the association of detailed Se pathway gene variation with cancer risk in 1420 CRC cases and 1421 controls within the European Prospective Investigation into Cancer and Nutrition (EPIC) cohort. We previously reported in a subset of this nested cohort with 966 case-control pairs that a higher Se status (ascertained by serum levels of Se and SELENOP) was associated with a lower CRC risk [5]. In these Western European subjects, the mean Se and SELENOP circulating levels were 84.0 μg/L and 4.3 mg/L in cases and 85.6 μg/L and 4.4 mg/L in controls, respectively. Thus, our present study was conducted in a generally suboptimal Se status population, as these Se concentrations are insufficient for optimal GPX3 expression and SELENOP saturation [5,6]. We now report the interaction between these genes and their corresponding pathways with Se status biomarkers and CRC risk.

2. Materials and Methods

2.1. Study Population and Design

EPIC is a multicenter prospective cohort study designed to investigate the association between diet, lifestyle, genetic and environmental factors and the incidence of cancers. The rationale and methods of the EPIC design have been described previously [24,25]. Briefly, 521,448 men and women mostly aged 25–70 years were enrolled between 1992–2000 in 23 sub-cohorts in 10 European countries (Denmark, France, Germany, Greece, Italy, The Netherlands, Norway, Spain, Sweden, and United Kingdom). The present analysis is based on participant data from all sub-cohorts except for Norway. At recruitment, standardized dietary, lifestyle and socio-demographic questionnaires including information on physical activity, education, smoking and medical history; anthropometric data, and blood samples were collected from participants. Blood and DNA samples are stored at the International Agency for Research on Cancer (IARC, Lyon, France) at −196 °C under liquid nitrogen for all countries except Denmark (−150 °C, nitrogen vapor) and Sweden (−80 °C freezers). Sample storage standardization including DNA extraction and quantification protocols were previously described in [26].

All study participants provided written informed consent. Ethical approval for the EPIC study was obtained from the review boards of the IARC (IARC Ethics Committee) and local participating centers. Study design methods were performed in accordance with the STROBE (Strengthening the Reporting of Observational Studies in Epidemiology) guidelines (https://www.strobe-statement.org/index.php?id=strobe-home).

2.2. Follow-Up for Cancer Incidence

Cancer incidence was determined through record linkage with population-based cancer registries (Denmark, Italy, Netherlands, Spain, Sweden, United Kingdom) or via a combination of methods, including the use of health insurance records, cancer and pathology registries, and active contact of study subjects or next-of-kin (France, Germany, Greece). Complete follow-up censoring dates for this study varied among centers, ranging between June 2002 and June 2003.

2.3. Selection of Cases and Controls and Study Design

Case subjects were men and women who developed first incident CRC after recruitment and before the latest follow-up date. Cancer incidence data were coded using the 10th Revision of the International Classification of Diseases (ICD-10) and the second revision of the International Classification of Disease for Oncology (ICDO-2). Colon cancers were defined as tumors in the cecum, appendix, ascending colon, hepatic flexure, transverse colon, splenic flexure, descending and sigmoid colon (C18.0-C18.7), and overlapping or unspecified origin tumors (C18.8 and C18.9). Rectal cancers were defined as tumors occurring at the recto-sigmoid junction (C19) or rectum (C20). Anal canal cancers (C21) were excluded. Colorectal cancer is the combination of the colon and rectal cancer cases.

All subjects with prior cancer diagnosis at any site (except non-melanoma skin cancer) were excluded. Cases were matched 1:1 by study center of enrollment, sex, age at blood collection, time of blood collection and fasting status, and menopausal status among women. Premenopausal women were matched on phase of menstrual cycle and postmenopausal women were matched on current hormonal therapy (HT) use. The matching was done as part of a previously published study on Se status [5], except for cases from Denmark for which new control subjects were identified due to problems with accessing the biobank. Furthermore, additional newly identified cases with their matching controls were also included for the genotyping from all participating countries but did not have biomarkers of Se status. Sweden was the only country of the nine participating in the genetic analysis for which we had no Se status data. Hence, there were 1478 cases and 1478 controls available for genotyping but the Se status information was only available for 966 of the cases and for 966 of the controls.

2.4. Gene Selection and Rationale

To examine selenoprotein gene and wider Se pathway gene variations in relation to CRC risk, we selected 1264 functional and haplotype tagging SNPs (tagSNPs) to comprehensively analyze common SNP variation in 164 Se pathway genes, which we assigned into eight functional pathways (listed in Supplementary Table S1). These included 42 genes in the primary selenoprotein pathway 1 (25 selenoprotein genes and 17 genes involved in Se transport and metabolism), and 122 genes in pathways 2–8 from (i) pathways affected by Se intake (Wnt, mTOR, Nrf2 and NF-κB signaling, endoplasmic reticulum and oxidative stress responses), and (ii) associated pathways of inflammatory response, apoptosis, DNA repair, Transforming growth factor (TGF) beta-signaling, and cell-cycle control [12] as detailed in Méplan and Hesketh, 2012 [13]. Variants in several genes from these affiliated pathways have been associated with CRC risk including regions of the Wingless/Integrated (Wnt) signaling gene *C-MYC* in CRC genome-wide association studies (GWAS) [27]. Our SNP analysis set was substantially enlarged from and included the 384 Se pathway SNPs (in 72 Se related genes) Se 'SNP-Chip' devised for a similar study of gene-Se interaction in a prostate cancer study within EPIC [28].

2.5. Tagging Single Nucleotide Polymorphism (tagSNP) Selection Protocol

A list of SNPs in all gene regions was compiled using the data from HapMap (release 27, based on dbSNP version b126 and NCBI genome build 36). TagSNPs were selected by use of the Tagger algorithm as implemented in the Haploview 3.2 software (Broad Institute, Cambridge, MA, USA). Parameters used for SNP selection were a Minor Allele Frequency (MAF) ≥5% in Caucasians and pairwise tagging ($r^2 \geq 0.8$). To include SNPs in promoter and potential regulatory regions, +/− 2 to 5 kilo base-pairs beyond the 5' and 3' ends were included. Additionally, known functional variants in our selected genes were added to the tagSNP list, e.g., for the selenoproteins these included rs7579, rs297299, and rs3877899 in *SELENOP* [4]. Selected SNPs were then assessed for suitability for the *Illumina GoldenGate*[TM] (Saffron Walden, Essex, UK) genotyping platform using Illumina's custom assay building platform (https://www.illumina.com/Documents/products/technotes/technote_goldengate_design.pdf).

Fifty-five SNPs which failed assay development criteria were replaced by proxy SNPs, i.e., those within the same genic region in high LD ($r^2 > 0.8$) to the original SNP. Proxy SNP replacements for functional selenoprotein SNPs which failed assay design included rs1800668 for *GPX1*-rs1050450, and rs5845 plus rs540049 for *SELENOF*-rs5859. However, there were no adequate proxies for *SELENOS*-rs34713741 or *GPX4*-rs713041.

2.6. Genotyping

A total of 1264 SNPs from 164 Se pathway genes were genotyped by *Illumina Goldengate*^{TM} in DNA samples available for 1478 case-control pairs matched within EPIC. Genotyping was performed simultaneously for cases and controls, blinded to case-control status (but with matched pairs analyzed in the same batch). A total of 62 replicate samples were genotyped to test for internal quality control, approximately 2 per genotyping plate, with the lowest reproducibility frequency for each of the replicates of 0.98. Samples with unclear or failed genotype calls were excluded from the analysis, leaving 1420 cases and 1421 controls for subsequent analyses.

From the 1264 initially selected, 96 SNPs failed genotyping, 27 failed Hardy–Weinberg Equilibrium (HWE), and 101 had less than 80% successfully genotyped samples. Thus, 1040 SNPs in 154 Genes (24 selenoprotein genes analyzed of 25, and 130 other Se pathways genes) with at least 80% genotypes across all genotyped samples were included in the final dataset (with a final genotyping call rate of 0.97, excluding zero call rate and those removed). Supplementary Table S2 provides the full gene and SNP list successfully analyzed in the current study.

2.7. Selenium Status Assays

Measurements of serum Se and SELENOP were previously done for a subset (966 cases and 966 controls) of the current analyzed cohort. The methods used were described in Hughes et al., 2015 and Hybsier et al., 2017 [5,29]. Briefly, total Se levels were measured in 4 uL of each serum sample using a bench-top total reflection X-ray fluorescence (TXRF) spectrometer (Picofox^{TM} S2, Bruker Nano GmbH, Berlin, Germany). SELENOP protein concentrations were ascertained from 20 μL of each serum sample by a colorimetric enzyme-linked immunoassay (Selenotest™, ICI GmbH, Berlin, Germany). For quality-control, the sample type (case or control) was blinded and two serum samples of known Se and SELENOP concentrations for intra-assay variability were included in each analysis plate. The samples were measured in duplicate and the mean concentration values, standard deviation (SD), and coefficient of variation (CV) were calculated. Duplicate samples with variances in concentration over 10% were re-measured. The evaluation was performed using GraphPad Prism 6.01 (GraphPad Software, La Jolla, CA, USA) using a four-parameter logistic function. The CV was 7.3% and 7.2% for controls 1 (SELENOP: 1.5 mg/L) and 2 (SELENOP: 8.6 mg/L), respectively.

2.8. Statistical Analysis

Both unconditional and conditional logistic regression analysis were carried out to assess the association of individual SNPs with CRC risk, adjusting for age (as a continuous variable), sex, and study center and provided similar results. We present the data for the unconditional logistic regression. Four standard genetic analysis models were tested for disease penetrance: multiplicative, additive, common recessive, and common dominant models [30]. Sub-group analyses by sex and anatomical sub-site of the colorectum (colon and rectum) were conducted. The associations between Se and SELENOP concentrations and genetic variants (coded as 0, 1, 2 corresponding to the number of minor alleles) were assessed among controls using linear regression models adjusted for age, sex, and center. Further adjustment by body mass index (BMI), smoking status, and physical activity did not change the results substantially.

Multiple testing corrections were performed by the Benjamini–Hochberg (BH) procedure [31]. *P*-values were also adjusted for correlated tests (P_{ACT}) to take account of the correlated nature of the

SNP data in biologically relevant and related pathways [32]. BH was performed for all SNPs, followed by P_{ACT} for the genes that had SNPs with $P < 0.01$.

We further employed exploratory gene- and pathway-based testing based on overall SNP variation to help identify possible important Se related biological pathways and genes with multiple risk variants that may be discounted in multiple testing corrections for the large number of SNPs with small effect sizes in a SNP by SNP approach. Genes were classified a priori into a primary best-known functional pathway based on the literature (listed in Supplementary Table S1). Gene- and pathway-based P-values were computed using the PIGE (Self-Contained Gene Set Analysis for Gene- and Pathway-Environment Interaction Analysis) R package which implements the modified Adaptive Rank Truncated Product (ARTP) test using a permutation algorithm [33] to accommodate gene-environment interactions (https://cran.rproject.org/web/packages/PIGE/index.html). Prior to this analysis, SNPs in high linkage disequilibrium (LD) were removed using AdaJoint [34] and the online tool SNPsnap (https://data.broadinstitute.org/mpg/snpsnap/about.html) so that all SNP pairs had LD $r^2 < 0.8$. Gene x Se status interactions were also examined using the PIGE R package. Although these methods do not identify individual susceptibility loci, they may help to identify a pathway that could modify the association between Se status and CRC risk. An advantage is that they do not require a priori knowledge of directionality for the variants.

All statistical tests were two-sided, and P-values < 0.05 were considered statistically significant (except $P < 0.1$ for P_{ACT}). Analyses were conducted using SAS version 9.2 (SAS Institute, Cary, NC, USA) and R (R Foundation for Statistical Computing, Vienna, Austria; http://www.R-project.org/) statistical packages.

3. Results

3.1. Baseline Characteristics of Participants

The baseline characteristics of participants are presented in Table 1. Colon and rectal cancer cases were diagnosed, on average, 4.1 and 4.2 years after blood collection, respectively. CRC cases were overall less likely to be physically active compared to controls. There were no data on Se supplement use for our study participants.

Table 1. Selected baseline characteristics of incident colon and rectal cancer cases and controls, the European Prospective Investigation into Cancer and Nutrition (EPIC) study, 1992–2003.

Characteristic	Colon Cancer Cases		Rectal Cancer Cases		Controls	
N	900		520		1419	
Women, N (%)	475	(52.8)	230	(44.2)	701	(49.4)
Mean age at blood collection, (SD) yrs	58.8	(7.5)	58.0	(6.9)	58.6	(7.4)
Mean years of follow-up (SD) yrs	4.1	(2.3)	4.2	(2.2)		
Smoking status, N (%) *						
Never	385	(42.8)	195	(37.5)	594	(41.9)
Former	299	(33.2)	177	(34)	460	(32.4)
Smoker	204	(22.7)	142	(27.3)	349	(24.6)
Physical activity, N (%) *						
Inactive	129	(14.3)	73	(14)	183	(12.9)
Moderately inactive	257	(28.6)	145	(27.9)	367	(25.9)
Moderately active	374	(41.6)	209	(40.2)	612	(43.1)
Active	75	(8.3)	55	(10.6)	148	(10.4)
BMI, kg/m² , (SD)	26.9	(4.36)	26.6	(3.92)	26.3	(3.84)

Table 1. *Cont.*

Characteristic	Colon Cancer Cases		Rectal Cancer Cases		Controls	
Country, N (%)						
Sweden	55	(6.1)	33	(6.3)	86	(6.1)
Denmark	174	(19.3)	164	(31.5)	340	(24)
The Netherlands	99	(11)	54	(10.4)	158	(11.1)
United Kingdom	166	(18.4)	74	(14.2)	250	(17.6)
Germany	110	(12.2)	69	(13.3)	169	(11.9)
France	22	(2.4)	6	(1.2)	29	(2)
Italy	144	(16)	58	(11.2)	198	(14)
Spain	101	(11.2)	45	(8.7)	141	(9.9)
Greece	29	(3.2)	17	(3.3)	48	(3.4)

* Percentages do not add up to 100% due to missing values. Abbreviations: BMI, body mass index; N, sample size; SD, standard deviation; yrs, years.

3.2. Se Pathway Genetic Variation and Colorectal Cancer (CRC) Risk Association

The 1040 tagging SNPs successfully analyzed from 154 genes and in HWE are shown in Supplementary Table S2 (which also provides all the genetic analysis results for CRC, plus stratified analyses for colon and rectal sub-site and by sex). These include 325 SNPs from 41 selenoprotein and Se transport/selenoprotein biosynthesis genes (designated as the primary Se pathway 1), and 715 variants from the other 113 wider Se metabolic pathway genes (pathways 2–8). A summary of the genetic associations before and after multiple testing corrections is provided in Supplementary Figure S1. Prior to adjustment for multiple comparisons, there were 144 SNPs in 63 genes nominally associated with CRC risk ($P < 0.05$ in at least one of the disease penetrance models tested; listed in Supplementary Table S3). There were 28 unique SNPs in LD with other associated SNPs and these are listed and highlighted in Supplementary Table S3 (tab 'LD CEU'). Among the 40 SNPs significantly associated with CRC risk from pathway 1, approximately half (21) were in 12 selenoprotein genes (i.e., 50% of the 24 selenoprotein genes successfully genotyped out of 25) and have the potential to affect the function or expression of individual selenoproteins, although this remains to be investigated. These 12 selenoprotein genes include those previously found associated with CRC risk (*GPX1, GPX4, SELENOF, TXNRD1, TXNRD2, TXNRD3*; for reviews, see [4,21]) and several with limited prior or no previous evidence of association with CRC risk (*DI01, GPX6, SELENOM, SELENON, SELENOT, SELENOV*). The other 19 SNPs associated with CRC in pathway 1 are in 8 of the 17 (47%) other Se transport/selenoprotein biosynthesis genes. Therefore, they have the potential to affect the synthesis of most selenoproteins (which also needs to be examined). Notably, 31% of the genes harboring SNPs associated with CRC risk (20 of 63) were related to selenoprotein biosynthesis and function implicated in protection from cancer development [4,21] with pathway 1 and 2 proteins involved in (1) Se homeostasis (*SELENOP, SEPHS1, SEPSEC, EFSEC, SCLY*), (2) antioxidant enzymes (*GPXs, TXNRDs, SELENON*), and (3) endoplasmic reticulum (ER) function or stress (*SELENOF, SELENOM, SELENOT*, and again *SELENON*). Additionally, several of these genes (e.g., *GPX1, GPX5, LRP2, SEPHS1, SELENOM, SELENON, TXNRD1*, and *TXNRD2*) had multiple SNPs and/or SNPs with raw *P*-values < 0.01 associated with CRC risk further supporting a role of selenoproteins, selenoprotein metabolism, ER stress, and oxidative stress in CRC development. Table 2 lists the SNPs in the primary Se pathway 1 with raw *P*-values < 0.01 associated with CRC risk.

In pathways 3–8, considering genes with multiple SNPs associated with CRC risk or SNPs with raw *P*-values < 0.01 for at least one genetic model, there were several notable and some novel associations with CRC risk for genes in pathways 3 (*C-MYC, FRZB*), 4 (*APAF1, BAX, FOXO3*), 5 (*IL12B, RPS6KA2, TRL4*), 6 (*MSH2, MSH3*) and 7 (*BMP2, BMPR2, SMAD3, SMAD7, TFGB1*).

Table 2. Single nucleotide polymorphisms (SNPs) associated with colorectal cancer (CRC) risk in primary selenium pathway 1 (selenium and selenoprotein transport, biosynthesis and metabolism) with raw *P*-values < 0.01 in at least one genetic model prior to multiple testing adjustment, the EPIC study, 1992–2003.

Gene/SNP/Genotype	CRC	Control	OR (95% CI)	P	P_{BH} +
GPX1/rs17080528					
GG	700	620	1.00 (ref)	0.010	0.703
GA	580	636	0.81 (0.69,0.95)		
AA	131	154	0.75 (0.58,0.97)		
Additive *	1411	1410	0.84 (0.75,0.95)	0.003	0.554
Dominant (GA + AA vs. GG)	1411	1410	0.80 (0.69,0.92)	0.003	0.534
Recessive (AA vs. GG + GA)	1411	1410	0.83 (0.65,1.06)	0.137	0.854
SELENOM/rs11705137					
AA	367	346	1.00 (ref)	0.024	0.753
AG	631	648	0.91 (0.75,1.09)		
GG	288	359	0.74 (0.60,0.92)		
Additive *	1286	1353	0.86 (0.78,0.96)	0.008	0.684
Dominant (AG + GG vs. AA)	1286	1353	0.85 (0.71,1.01)	0.064	0.815
Recessive (GG vs. AA + AG)	1286	1353	0.79 (0.66,0.95)	0.012	0.710
SELENON/rs4659382					
GG	783	713	1.00 (ref)	0.019	0.747
GC	509	573	0.80 (0.69,0.94)		
CC	96	107	0.82 (0.61,1.10)		
Additive *	1388	1393	0.86 (0.76,0.97)	0.011	0.710
Dominant (GC + CC vs. GG)	1388	1393	0.81 (0.69,0.94)	0.005	0.625
Recessive (CC vs. GG + GC)	1388	1393	0.90 (0.67,1.20)	0.455	0.954
SEPHS1/rs2275129					
GG	361	423	1.00 (ref)	0.032	0.780
GC	726	690	1.23 (1.03,1.47)		
CC	321	295	1.28 (1.04,1.58)		
Additive *	1408	1408	1.14 (1.02,1.26)	0.017	0.747
Dominant (GC + CC vs. GG)	1408	1408	1.25 (1.05,1.47)	0.010	0.697
Recessive (CC vs. GG + GC)	1408	1408	1.12 (0.94,1.34)	0.217	0.885
TXNRD1/rs11111979 ^					
GG	395	429	1.00 (ref)	0.015	0.745
GC	627	680	1.00 (0.84,1.20)		
CC	279	230	1.34 (1.07,1.67)		
Additive *	1301	1339	1.14 (1.02,1.27)	0.022	0.749
Dominant (GC + CC vs. GG)	1301	1339	1.09 (0.92,1.28)	0.315	0.932
Recessive (CC vs. GG + GC)	1301	1339	1.33 (1.10,1.62)	0.004	0.566

+ = After Benjamini–Hochberg (BH) multiple testing correction; * = Additive models impose a structure in which each additional copy of the variant allele increases the response (log odds ratio) by the same amount; ^ = *TXNRD1* rs11111979 was borderline significant after adjustment for correlated tests (P_{ACT} = 0.10). EPIC = European Prospective Investigation into Cancer and Nutrition.

None of the SNPs in the primary Se pathway 1 remained significant after multiple testing corrections by the BH procedure. Overall, only 6 SNPs harbored by more distantly related genes in cell-signaling pathways retained significance (*FRZB*, *SMAD3*, and *SMAD7*; see Table 3). Genes harboring SNPs with raw *P*-values < 0.01 with CRC risk for at least one genetic model (21 genes/34 SNPs) were further considered for gene-wide variance significance by the P_{ACT} method. For pathway 1, the *TXNRD1* selenoprotein variant rs11111979, an intron 3'–5'UTR SNP previously associated with healthy aging [35], remained borderline significant for an association with CRC risk (P_{ACT} = 0.100; P_{ACT} significance threshold was $P < 0.1$) in the recessive genetic model. Including *FRZB*, *SMAD3* and

SMAD7, the other wider pathway genes retaining significance were *C-MYC* (P_{ACT} = 0.032), *BMP2* (P_{ACT} = 0.012) and *BAX* (P_{ACT} = 0.035).

Table 3. Single Nucleotide Polymorphisms (SNPs) statistically significantly associated with colorectal cancer (CRC) risk after Benjamini–Hochberg (BH) multiple testing correction, the EPIC study, 1992–2003.

Gene/SNP/Genotype	CRC	Control	OR (95% CI)	P	P_{BH}
FRZB/rs17265803 ^					
AA	844	976	1.00 (ref)	3.04E-06	0.003
AG	315	240	1.56 (1.28,1.89)		
Additive *	1163	1232	1.35 (1.13,1.62)	0.001	0.372
Dominant (AG + GG vs. AA)	1163	1232	1.48 (1.22,1.79)	6.77E-05	0.034
Recessive (GG vs. AA + AG)	1163	1232	0.26 (0.09,0.80)	0.018	0.747
SMAD3/rs7180244 ^					
GG	994	1183	1.00 (ref)	2.22E-16	1.12E-12
GC	372	198	2.33 (1.92,2.83)		
Additive *	1372	1388	2.16 (1.79,2.60)	1.11E-15	3.74E-12
Dominant (GC + CC vs. GG)	1372	1388	2.28 (1.88,2.77)	1.11E-16	1.12E-12
Recessive (CC vs. GG + GC)	1372	1388	0.87 (0.29,2.61)	0.804	0.997
SMAD7/rs11874392					
AA	478	400	1.00 (ref)	1.39E-07	2.82E-04
AT	704	671	0.88 (0.74,1.04)		
TT	222	337	0.55 (0.44,0.68)		
Additive *	1404	1408	0.75 (0.68,0.84)	1.99E-07	3.36E-04
Dominant (AT + TT vs. AA)	1404	1408	0.77 (0.65,0.90)	0.001	0.372
Recessive (TT vs. AA + AT)	1404	1408	0.59 (0.49,0.72)	6.26E-08	1.58E-04
SMAD7/rs12953717					
GG	366	470	1.00 (ref)	3.47E-05	0.019
GA	671	643	1.36 (1.14,1.62)		
AA	335	273	1.60 (1.29,1.98)		
Additive *	1372	1386	1.27 (1.14,1.41)	9.07E-06	0.006
Dominant (GA + AA vs. GG)	1372	1386	1.43 (1.21,1.68)	2.38E-05	0.014
Recessive (AA vs. GG + GA)	1372	1386	1.32 (1.10,1.59)	0.003	0.534
SMAD7/rs4939827					
AA	433	378	1.00 (ref)	6.46E-06	0.005
AG	664	634	0.92 (0.77,1.09)		
GG	248	357	0.60 (0.49,0.75)		
Additive *	1345	1369	0.79 (0.71,0.87)	9.46E-06	0.006
Dominant (AG + GG vs. AA)	1345	1369	0.80 (0.68,0.95)	0.009	0.697
Recessive (GG vs. AA + AG)	1345	1369	0.64 (0.53,0.77)	1.66E-06	0.002
SMAD7/rs6507874					
AA	467	389	1.00 (ref)	1.93E-06	0.002
AG	705	677	0.87 (0.73,1.03)		
GG	234	336	0.57 (0.46,0.71)		
Additive *	1406	1402	0.77 (0.69,0.86)	1.26E-06	0.002
Dominant (AG + GG vs. AA)	1406	1402	0.77 (0.65,0.90)	0.001	0.420
Recessive (GG vs. AA + AG)	1406	1402	0.63 (0.52,0.76)	1.16E-06	0.002

^ = Results for the rare homozygous genotypes are omitted for these SNPs due to the small sample numbers with these genotypes; * = Additive models impose a structure in which each additional copy of the variant allele increases the response (log odds ratio) by the same amount. EPIC = European Prospective Investigation into Cancer and Nutrition.

Supplementary Table S3 also catalogs the SNPs with raw significant *P*-values and the BH corrections stratified by cancer sub-site (comprising 138 SNPs in 65 genes for colon in tab '*colon cancer*' and 123 SNPs in 54 genes for rectum listed in tab '*rectal cancer*'). Additionally, the tab '*All*' lists all the SNPs showing an association for CRC, colon only, rectal only, plus the analyses stratified by sex. Generally, there was predominate overlap in the genes associated with CRC and sub-site risks. Genes containing variants uniquely associated only with sub-site risk plus raw *P*-values < 0.01 comprised rs12124257 in *PTGS2* for colon cancer and 4 SNPs in *IL10* for rectal cancer.

3.3. Associations Between Se Pathway Genetic Variation and Se Status

Among controls, 99 different SNPs in 55 genes were nominally associated with Se status (raw *P*-values < 0.05). From these 99 variants, 87 SNPs from 45 genes were associated with either Se or SELENOP levels (55 SNPs in 33 genes and 56 SNPs in 37 genes, respectively) while the other 12 variants from 10 genes were associated with both Se status measures, including 2 each in *HIF1A* and *SMAC*. Eight pathway 1 genes harbored 14 SNPs significant for Se level changes (including 7 SNPs in 5 selenoprotein genes) while 9 pathway 1 genes carried 15 variants significant for SELENOP status association (11 SNPs in 6 selenoprotein genes). However, none of the associations retained significance after BH multiple testing adjustments. These SNP IDs together with the beta coefficients for change in Se (μg/L) or SELENOP (mg/L) are listed according to gene pathway in Supplementary Table S4.

3.4. Pathway Analysis

An exploratory analysis of CRC risk with gene variation and gene x Se status interaction within eight predefined pathways was performed using the PIGE package. A summary of the main PIGE results per pathway is presented in Table 4, while Supplementary Table S5 provides all the *P*-values for each gene per pathway designation. Considering nominal significance for association with disease risk by pathway of *P* < 0.05, then these analyses suggest that TGF-beta signaling (*P* < 0.001) is the sole pathway highly associated with CRC risk independent of Se status interaction. Antioxidant/redox pathway genetic variation combined with Se status interactions was associated with a significant effect on CRC risk (*P* = 0.011 and 0.010 for Se and SELENOP interactions, respectively), possibly driven by SNPs in *HIF1A*, *KEAP1*, *GPX7*, *CAT*, and *SOD2* (when considering the *P*-values for each individual gene regarding gene only variation and gene x Se status interaction; see Supplementary Table S5). In contrast, the risk association with gene variation in the apoptosis pathway seems to depend more on interaction with Se levels (*P* = 0.003) but not with SELENOP concentrations (*P* = 0.105). For gene only analyses there were several genes across the pathways associated with CRC risk including the pathway 1 genes *GPX1*, *SELENOM*, *SELENON*, and *SEPHS1*, from which only *SELENON* was significant for both the gene only and gene x Se interaction PIGE analyses (Supplementary Table S5). In agreement with previous large gene association and GWAS studies, overall genetic variation in the *SMAD3*, *SMAD7*, *BMP2*, and *BMPR2* genes was associated with CRC risk [36–38]. Excluding individuals with no measurements of Se or SELENOP concentrations did not substantially change the main gene only PIGE results (both sets of results are provided in Supplementary Table S5).

Table 4. *P*-values for genetic pathways and pathway-selenium (Se) status interactions and colorectal cancer risk, the EPIC study, 1992–2003.

Pathway	$P_{\text{Pathway Only}}$	$P_{\text{Pathway Only (non-Missing Se Status)}}$ ***	$P_{\text{Pathway x Se Interaction}}$	$P_{\text{Pathway x SELENOP Interaction}}$
Se and Selenoproteins *	0.217	0.098	0.615	0.726
Antioxidant and Redox	0.173	0.072	0.011	0.010
Cell signaling **	0.307	0.489	0.223	0.872
Apoptosis	0.361	0.097	0.003	0.105
Inflammation	0.822	0.262	0.199	0.607
DNA repair	0.739	0.432	0.175	0.088
TGFβ signaling	<0.001	0.001	0.061	0.764
Cell cycle control	0.398	0.475	0.097	0.449

* Se and selenoprotein transport, biosynthesis & metabolism. ** Includes Wnt, mTOR, NfkB, and Nrf2 signaling.
*** Includes only participants with non-missing blood Se or SELENOP concentrations. EPIC = European Prospective Investigation into Cancer and Nutrition.

4. Discussion

The results of this prospective nested case-control study represent the largest reported analysis of both the association of Se pathway SNP variation and the interaction with Se status biomarkers (serum Se levels and SELENOP protein concentrations) with CRC risk. The analysis of 1,040 tagSNPs in 154 Se pathway genes in DNA samples from 1,420 CRC cases and 1,421 controls within EPIC indicated that 144 of these SNPs in 63 genes were nominally associated with CRC risk. However, for pathway 1 only the *TXNRD1* selenoprotein gene rs11111979 SNP retained borderline significance after correction for multiple testing. For pathways 2–8, variants in *BAX, BMP2, C-MYC, FRZB, SMAD3,* and *SMAD7* passed significance thresholds following these adjustments.

Selenoprotein genes nominally associated with CRC risk included several with limited or no prior evidence (*DIO1, GPX6, SELENOM, SELENON, SELENOT, SELENOV*) and those reported in several studies (*GPX1, GPX4, SELENOF, TXNRD1, TXNRD2, TXNRD3*) for an association with CRC (or specifically colon or rectal cancer) risk. This latter group of genes has been more extensively examined due to their putative roles related to cancer prevention in colonic tissue (for reviews, see [4,21]), while the former group of selenoprotein genes have generally less well-characterized function, especially regarding how they may affect colorectal function and CRC development. Overall, any functional consequences from genetic variations in these genes, together with Se status, may affect several oxidative stress, inflammatory, and signal translation pathways implicated in colorectal carcinogenesis [13,39]. Notably several of these genes are ER-resident selenoproteins (*SELENOF, SELENOM, SELENON, SELENOT*), thought to be involved in ER-stress response and calcium flux, comprising a potentially important mechanism of selenoprotein-related cancer prevention or promotion [40].

None of the 3 *GPX1* SNPs (rs17080528, rs3448, rs9818758) or rs2074451 in *GPX4* associated with CRC risk are in high LD (i.e., $r^2 \geq 0.8$) to the functional *GPX1* Pro/Leu rs1050450 and *GPX4* rs713041 SNPs (for which the *Illumina* assays failed) previously implicated in prostate, breast, lung (rs1050450), and CRC risk (rs713041) [4]. However, from these pathway 1 genes, only a *TXNRD1* selenoprotein variant (rs11111979), one of the three thioredoxin reductases which function in redox control [8], remained borderline significant for an increased CRC risk when applying gene-wide variance considerations by the P_{ACT} method. Interestingly, this SNP inducing a change in the 5′untranslated region of *TXNRD1*, among others in *TXNRD1*, was previously observed to be associated with age-related physical performance [35], and age is a primary risk factor for CRC development [41]. In the wider metabolic pathway, following adjustment for multiple testing, genotypes for SNPs in Wnt, TGF-beta signaling, and apoptosis pathway genes (*C-MYC, FRZB, SMAD3, SMAD7, BMP2,* and *BAX*) were also significantly associated with CRC risk.

Positive associations of selenoprotein gene variants with CRC risk have been more commonly reported in areas with suboptimal Se availability such as European populations, than regions with generally adequate Se intake (e.g., North America). However, tagSNPs in several selenoprotein genes (*GPX3, TXNRD3, SELENON, SELENOF,* and *SELENOX*) were also associated with colon or rectal cancer risk and/or survival outcomes in two separate studies of several large case-control USA cohorts drawn from populations with generally adequate dietary Se intakes [22,23]. Associations of multiple SNPs in the same selenoprotein gene with CRC risk were observed in this study for *GPX1, SELENON, TXNRD1,* and *TXNRD2* (3, 4, 3, and 3 SNPs, respectively), broadly comparable to previous reports [4]. As reported by Slattery et al. in 2012 many of the same selenoprotein genes were separately associated with colon and rectal cancer risk in sub-site analyses although risks often differed by SNP [22]. From the 4 *SELENON* variants that were associated with CRC risk (rs11247735, rs2072749, rs4659382, and rs11247710), the first 3 were previously associated with rectal cancer risk in this North American cohort [22]. In our sub-site analyses, rs11247735, rs2072749, and rs11247710 were associated with rectal cancer risk only, and rs4659382 with both colon and rectal cancer. We also found further modifications of gene only and gene-Se risk for CRC by sex, as indicated by our previous studies of selenoprotein genetic variation in a Czech population [19] and Se status in the EPIC study [5]. This reflects the importance of interactions between Se intake, Se status and genotype, sex and CRC sub-type risks (reviewed in [4]).

Prior to this study there were few data available on the interaction of selenoprotein genotype and Se status regarding CRC risk, apart from a study in a Se replete population of North American women which reported that the null results for serum Se did not differ by selenoenzyme (*GPX1-4* and *SELENOP*) genetic variants [17]. The effect of Se pathway SNPs on the efficacy of Se utilization may be particularly relevant to CRC risk in populations with sub-optimal Se status, such as this study within EPIC [5]. We observed that numerous genetic variations were associated with Se status levels (as assessed by serum Se and SELENOP concentration), although these were not significant after adjustment for multiple tests. However, in the PIGE analysis overall gene and pathway genetic variation interacted with biomarkers of Se status to alter CRC risk. As expected, there were several variations in pathway 1 nominally associated with Se status levels. These included selenoprotein genes *SELENON*, *SELENOP*, *SELENOS*, and *TXNRD1* that are regulated by Se availability and whose genetic variations have been previously shown to affect blood and tissue Se levels [6,21,42,43]. Transgenic mouse studies underlined the critical function of SELENOP for Se organification and transport [11]. Thus, SNP interactions with SELENOP levels may be particularly important regarding CRC risk as serum SELENOP is a functional marker of Se status and is more associated with CRC risk than Se in this cohort [5]. Notably, all 4 of the variants associated with both SELENOP status and CRC risk in the SNP only analysis were from 2 selenoprotein genes; rs4659382, rs11247710, and rs2072749 in *SELENON* and the P_{ACT} borderline significant rs11111979 variant in *TXNRD1*. Selenoproteins SELENOP, SELENON, and TXNRD1 are antioxidant enzymes and their genetic variations plus regulation by SELENOP levels may be important factors in relation to colorectal carcinogenesis. We observed an association of SELENOP levels with rs6413428 in *SELENOP*, which in a SNP-only analysis was previously observed to be associated with CRC risk in the USA [44], an area of generally high Se status, but not in our study. Another *SELENOP* variant, rs3877899, was also associated with SELENOP status, while the GG genotype for this SNP previously showed the highest significant correlation of all selenoprotein genotypes tested between serum Se and activity of the vital antioxidant enzyme thioredoxin reductase [45]. This latter study also showed a correlation between serum Se and increased DNA damage with *SELENOS*-rs4965373 under peroxide challenge. We selected rs12910524 in *SELENOS* as a proxy tagSNP for this variant (as rs4965373 failed *Illumina* assay development) and found that it was significantly associated with Se levels. The lipoprotein megalin receptor (LRP2) protein appears to mediate SELENOP uptake to various tissues and affect plasma Se status levels [46,47]. Here, the *LRP2* SNPs rs12614394, rs2229266, rs2389557, rs700552, and rs9789747 were associated with CRC risk alone while the rs3755166 promoter SNP was associated with Se levels. Previously, rs3755166 has been associated with Alzheimer's disease with the rare allele showing decreased transcriptional activity [48]. Intriguingly, this indicates a potential mechanism for the suggested link of sub-optimal Se status with neurodegenerative disease [49].

Supporting the data presented here, *GPX1* and *GPX4* selenoprotein gene loci have been implicated in GWAS of inflammatory bowel disease (IBD), which is a risk factor for CRC development [50,51]. Additionally, the rs7901303 variant from the selenophosphate synthetase 1 (*SEPHS1*) gene, which plays a major role in selenoprotein synthesis, was associated in this study with CRC risk (before multiple testing corrections). rs7901303 was previously associated with risk of Crohn's disease in interaction with serum Se levels in a sub-optimal Se population of New Zealand [52]. The genetic associations identified in these studies suggest therefore a key role of the corresponding proteins in colorectal function and/or the carcinogenic process.

Genomic studies and animal models have shown Se intake to not only affect expression of selenoprotein genes but also pathways key to colorectal carcinogenesis such as the antioxidant response, immune and inflammatory pathways (including NFkB and Nrf2 signaling) and the Wnt signaling pathway [4,13,53]. Furthermore, expression of constituents of these metabolic pathways has been shown to be affected by Se level in human rectal biopsies [54]. Therefore, in addition to a focus on Se metabolism and selenoprotein genes, the present analysis also encompassed a substantial examination of genetic variations in these selenium-relevant pathways. Associations of multiple SNPs in the same gene (several of which are novel) with CRC risk were observed in genes from pathways 2–8, e.g., *BAX, GPX5, FOXO3, IL12B, TLR4, MSH2, MSH3, TGFB1*, as well as *IL10* with rectal cancer risk. Polymorphisms in several of these genes have previously been associated with CRC risk [55–62]. After BH multiple testing corrections, SNPs in cell-signaling pathways retained significance (*FRZB, SMAD3,* and *SMAD7*). The variants in both *SMAD* genes were previously linked to CRC by GWAS, suggesting a role of these variants in CRC development [38]. The association of rs17265803 in *FRZB* appears to be novel and it is not in LD with the functional *FRZB* genetic variant Arg324Gly (rs7775) previously reported to be associated with an increased CRC risk [63], although this was not replicated in a nested case-control study [64]. Additional genetic variants retaining significance by P_{ACT} were rs6983267 in *C-MYC* previously identified in a meta-GWAS [65], rs235770 in *BMP2* previously associated with colon cancer risk [37], and a novel association of rs4645887 in *BAX*.

In the pathway analysis, all the Se pathway genes were grouped into a primary best-known functional pathway and were analyzed for the association of whole gene and whole pathway genetic variation with CRC risk, and in interaction with Se status. Neither gene only variation or interactions with Se status in the core pathway 1 selenoprotein and biosynthesis pathway were associated with CRC risk by pathway, although gene only variation for *GPX1, SELENOM, SELENON,* and *SEPHS1* plus gene x Se status interactions for *SELENON* (with SELENOP) and *PSTK* (with Se) were associated with CRC risk. In the gene x Se analysis, only pathway 2 (antioxidant/redox) was significant for an association with CRC risk for both Se and SELENOP. Alternatively, it also remains possible that the genetic 'noise' from any irrelevant selenoproteins masked the overall risk associations for pathway 1 (based on the rationale that most genes in pathways like oxidative stress are important in cancer prevention but that some of the selenoproteins may be irrelevant to colorectal carcinogenesis, as they are included solely because they share selenocysteine motifs). This is partly supported by the strongest association (by PIGE) in pathway 1 for gene variance in *GPX1*, which has been previously implicated in risk of various cancers [4]. However, these pathway divisions cannot reflect, for example, the biological overlaps with the antioxidant selenoprotein genes in pathway 1 and their non-Se containing counterparts in pathway 2. Several aspects of our data suggest a potentially under-appreciated focus on variation in apoptosis genes (pathway 4) and CRC risk that may also be modified by Se status. These comprise the association of several SNPs in both the *FOXO3* and *BAX* genes (including the P_{ACT} significant rs4645887 variant in *BAX*) with CRC risk, significance of *FOXO3* for overall gene variation, and several significant findings of SNP x Se and gene x Se status interactions for genes in this pathway (e.g., *SMAC, CASP8, MAPK8,* and *MAPK9*). Overall, our analyses suggest that genetic variation in TGF beta signaling (pathway 7), which includes members such as *BMP2, BMPR2, SMAD3,* and *SMAD7* implicated in CRC risk by previous large case-control and GWAS reports [36–38], is sufficient to alter CRC risk, independent of Se status interaction, while SNP risk associations attributed to the antioxidant and apoptosis pathways may be significantly modified by Se status interactions.

Strengths and weaknesses of our study design for the Se status analyses have been discussed earlier [5]. The hypothesis-driven approach and appreciable sample size within a large, prospective study allowed an extensive examination of Se pathway genetic variation (including gene pathway analyses) and the interaction with robust markers of Se status regarding CRC risk. Despite the large sample size, gene pathway, gene–Se interaction analysis and some stratified analyses had limited power, particularly analyses by sex and anatomical sub-sites. The pathway designations were assigned based on known function from the literature, and there will be interactions between these pathways that we were not able to model. Finally, as most of the reported associations involve tagSNPs of no known functionality (or the actual contributing functional variant(s) they tag) additional genetic mapping and lab-based studies will be needed to explore these aspects.

5. Conclusions

In summary, the present study indicates that genetic variation in selenoprotein genes and genes in antioxidant/redox, Wnt, apoptotic, and TGF-beta signaling pathways may modify risk of CRC development. Furthermore, for genes in antioxidant/redox and apoptotic pathways the influence of SNPs on the disease risk is also dependent on interaction with Se status. Overall, these results taken together with our previous study [5] suggest that risk of CRC may be modified by genotype, Se status, sex, and gene variation interactions within biological pathways. Thus, will individuals harboring these genotypes benefit from increased Se intake, including consideration of 'Se adequate' environments, such as the US, where Se intervention trials have not shown a significant benefit in the general population [66]? Before such a recommendation can be defined, further examination of these findings in other populations and investigation of Se metabolism is needed to clarify the relevance of the Se pathway and signaling genotypes for CRC etio-pathogenesis, especially for individuals with suboptimal Se status.

Supplementary Materials: The following are available online at http://www.mdpi.com/2072-6643/11/4/935/s1: Figure S1: Selenium pathway genetic analysis flowchart, Table S1: Se pathway SNP study: List of genes and the 8 pathway designations, the EPIC study, 1992–2003, Table S2: All Se pathway SNPs by gene and pathway and association with CRC risk, the EPIC study, 1992–2003, Table S3: All Se pathway SNPs (by gene and pathway) significantly associated with CRC risk before multiple testing corrections, the EPIC study, 1992–2003, Table S4: Se pathway SNPs from pathways 1–8 associated with Se or SELENOP Concentrations Among Controls, the EPIC study, 1992–2003, Table S5: Se pathway PIGE P values for each gene and CRC risk association per pathway designation, the EPIC study, 1992–2003.

Author Contributions: Conceptualization, D.J.H., V.F., C.M., J.E.H., L.S., M.J. and E.R.; Data curation, D.J.H., J.E.H., V.F., L.S. and M.J.; Formal analysis, V.F., D.J.H., J.S.J., W.Z., C.M. and M.J.; Funding acquisition, D.J.H., V.F., J.E.H., C.M., L.S. and M.J. Investigation, D.J.H., V.F., C.M., J.E.H., L.S., S.H. (Sandra Hybsier), A.S. and M.J.; Methodology, V.F., D.J.H., J.E.H., C.M., M.J., L.S. and E.R.; Project administration, D.J.H., V.F., C.M., J.E.H. and M.J. Resources, D.J.H., V.F., C.M., J.E.H., M.J., and all other EPIC co-authors (K.O., A.T. (Anne Tjønneland), H.O., V.P., M.-C.B.-R.,T.K., V.K., K.A., A.Tr., A.K. (Anna Karakatsani), A.K. (Anastasia Kotanidou), R.T., S.P., G.M., C.A., A.N., B.B.-d.-M., R.C.H.V., E.W., G.S., T.H.N., L.L.-B., J.R.Q., J.M.H., M.R.-B., A.B., B.G., S.H. (Sophia Harlid), K.E.B., N.W., K.-T.K., M.G., N.M., H.F., K.T., D.A., and E.R.); Software, V.F., D.J.H., A.S., J.S.J., W.Z., L.S. and M.J.; Supervision, D.J.H., V.F., L.S. and M.J.; Validation, D.J.H., V.F., C.M., A.S., S.H. (Sandra Hybsier) and L.S.; Writing—original draft, D.J.H., V.F., M.J., C.M., and J.E.H.; Writing—review and editing, all other EPIC co-authors (K.O., A.T. (Anne Tjønneland), H.O., V.P., M.-C.B.-R.,T.K., V.K., K.A., A.T. (Antonia Trichopoulou), A.K. (Anna Karakatsani), A.K. (Anastasia Kotanidou), R.T., S.P., G.M., C.A., A.N., B.B.-d.-M., R.C.H.V., E.W., G.S., T.H.N., L.L.-B., J.R.Q., J.M.H., M.R.-B., A.B., B.G., S.H. (Sophia Harlid), K.E.B., N.W., K.-T.K., M.G., N.M., H.F., K.T., D.A. and E.R.) reviewed and approved the manuscript and commented on the analysis and interpretation of the findings. The lead authors (D.J.H., V.F., C.M., J.E.H. and M.J.) affirm that the manuscript is an honest, accurate, and transparent account of the study being reported; that no important aspects of the study have been omitted; and that any discrepancies from the study as planned have been explained.

Funding: Funding for this study was provided by the Health Research Board of Ireland health research awards HRA_PHS/2013/397 and HRA_PHS/2015/1142 (principal investigator: D.J.H.). We thank all the members of EPIC for their work on the EPIC cohort study. The EPIC study was supported by "Europe Against Cancer" Programme of the European Commission (SANCO); Ligue contre le Cancer; Institut Gustave Roussy; Mutuelle Générale de l'Education Nationale; Institut National de la Santé et de la Recherche Médicale (INSERM); German Cancer Aid; German Cancer Research Center; German Federal Ministry of Education and Research; Danish Cancer Society; Health Research Fund (FIS) of the Spanish Ministry of Health; the CIBER en Epidemiología y Salud Pública (CIBERESP), Spain; ISCIII RETIC (RD06/0020); Spanish Regional Governments of Andalusia, Asturias, Basque Country, Murcia (No 6236) and Navarra and the Catalan Institute of Oncology; Cancer Research UK; Medical Research Council, UK; the Hellenic Health Foundation; Italian Association for Research on Cancer; Italian National Research Council; Compagnia di San Paolo; Dutch Ministry of Public Health, Welfare and Sports (VWS), Netherlands Cancer Registry (NKR), LK Research Funds, Dutch Prevention Funds, Dutch ZON (ZorgOnderzoek Nederland), World Cancer Research Fund (WCRF), Statistics Netherlands (The Netherlands); Swedish Cancer Society; Swedish Scientific Council; Regional Governments of Skane and Vasterbotten, Sweden; and Nordforsk center of excellence programme HELGA. L.S. was supported by Deutsche Forschungsgemeinschaft (DFG Research Unit 2558 TraceAge, Scho 849/6-1). D.P. was supported by the Associazione Italiana per la Ricercasul Cancro-AIRC-Italy.

Conflicts of Interest: The authors declare no competing interests. Funding support for the EPIC study is described in the acknowledgements; there were no financial relationships with any organizations that might have an interest in the submitted work in the previous three years, and no other relationships or activities that could appear to have influenced the submitted work. While L.S. is the founder of selenOmed GmbH, a company involved in improving Se diagnostics, the data were analyzed before the founding of selenOmed and the analyses were conducted completely blinded to any clinical finding. For information on how to apply for gaining access to EPIC data and/or biospecimens, please follow the instructions at http://epic.iarc.fr/access/index.php. IARC Disclosure Statement: Where authors are identified as personnel of the International Agency for Research on Cancer/World Health Organization, the authors alone are responsible for the views expressed in this article and they do not necessarily represent the decisions, policy or views of the International Agency for Research on Cancer/World Health Organization.

References

1. Ferlay, J.; Soerjomataram, I.; Dikshit, R.; Eser, S.; Mathers, C.; Rebelo, M.; Parkin, D.M.; Forman, D.; Bray, F. Cancer incidence and mortality worldwide: Sources, methods and major patterns in globocan 2012. *Int. J. Cancer* **2015**, *136*, E359–E386. [CrossRef] [PubMed]

2. Cappellani, A.; Zanghi, A.; Di Vita, M.; Cavallaro, A.; Piccolo, G.; Veroux, P.; Lo Menzo, E.; Cavallaro, V.; de Paoli, P.; Veroux, M.; et al. Strong correlation between diet and development of colorectal cancer. *Front. Biosci.* **2013**, *18*, 190–198.

3. Slattery, M.L.; Lundgreen, A.; Herrick, J.S.; Caan, B.J.; Potter, J.D.; Wolff, R.K. Diet and colorectal cancer: Analysis of a candidate pathway using SNPS, haplotypes, and multi-gene assessment. *Nutr. Cancer* **2011**, *63*, 1226–1234. [CrossRef] [PubMed]

4. Meplan, C.; Hesketh, J. Selenium and Cancer: A Story that Should not be Forgotten-Insights from Genomics. *Cancer Treat. Res.* **2014**, *159*, 145–166. [PubMed]

5. Hughes, D.J.; Fedirko, V.; Jenab, M.; Schomburg, L.; Meplan, C.; Freisling, H.; Bueno-de-Mesquita, H.B.; Hybsier, S.; Becker, N.P.; Czuban, M.; et al. Selenium status is associated with colorectal cancer risk in the European prospective investigation of cancer and nutrition cohort. *Int. J. Cancer* **2015**, *136*, 1149–1161. [CrossRef] [PubMed]

6. Combs, G.F., Jr. Biomarkers of selenium status. *Nutrients* **2015**, *7*, 2209–2236. [CrossRef]

7. Fairweather-Tait, S.J.; Bao, Y.; Broadley, M.R.; Collings, R.; Ford, D.; Hesketh, J.E.; Hurst, R. Selenium in human health and disease. *Antioxid. Redox Signal.* **2011**, *14*, 1337–1383. [CrossRef] [PubMed]

8. Labunskyy, V.M.; Hatfield, D.L.; Gladyshev, V.N. Selenoproteins: Molecular pathways and physiological roles. *Physiol. Rev.* **2014**, *94*, 739–777. [CrossRef]

9. Gladyshev, V.N.; Arnér, E.S.; Berry, M.J.; Brigelius-Flohé, R.; Bruford, E.A.; Burk, R.F.; Carlson, B.A.; Castellano, S.; Chavatte, L.; Conrad, M.; et al. Selenoprotein Gene Nomenclature. *J. Biol. Chem.* **2016**, *291*, 24036–24040.

10. Hatfield, D.L.; Tsuji, P.A.; Carlson, B.A.; Gladyshev, V.N. Selenium and selenocysteine: Roles in cancer, health, and development. *Trends Biochem. Sci.* **2014**, *39*, 112–120. [CrossRef] [PubMed]

11. Burk, R.F.; Hill, K.E. Regulation of Selenium Metabolism and Transport. *Annu. Rev. Nutr.* **2015**, *35*, 109–134. [CrossRef] [PubMed]

12. Hesketh, J.; Meplan, C. Transcriptomics and functional genetic polymorphisms as biomarkers of micronutrient function: Focus on selenium as an exemplar. *Proc. Nutr. Soc.* **2011**, *70*, 365–373. [CrossRef]

13. Meplan, C.; Hesketh, J. The influence of selenium and selenoprotein gene variants on colorectal cancer risk. *Mutagenesis* **2012**, *27*, 177–186. [CrossRef] [PubMed]

14. Irons, R.; Carlson, B.A.; Hatfield, D.L.; Davis, C.D. Both selenoproteins and low molecular weight selenocompounds reduce colon cancer risk in mice with genetically impaired selenoprotein expression. *J. Nutr.* **2006**, *136*, 1311–1317. [CrossRef]

15. Irons, R.; Tsuji, P.A.; Carlson, B.A.; Ouyang, P.; Yoo, M.H.; Xu, X.M.; Hatfield, D.L.; Gladyshev, V.N.; Davis, C.D. Deficiency in the 15-kDa selenoprotein inhibits tumorigenicity and metastasis of colon cancer cells. *Cancer Prev. Res.* **2010**, *3*, 630–639. [CrossRef]

16. Hu, Y.; McIntosh, G.H.; Le Leu, R.K.; Young, G.P. Selenium-enriched milk proteins and selenium yeast affect selenoprotein activity and expression differently in mouse colon. *Br. J. Nutr.* **2010**, *104*, 17–23. [CrossRef] [PubMed]

17. Takata, Y.; Kristal, A.R.; King, I.B.; Song, X.; Diamond, A.M.; Foster, C.B.; Hutter, C.M.; Hsu, L.; Duggan, D.J.; Langer, R.D.; et al. Serum selenium, genetic variation in selenoenzymes, and risk of colorectal cancer: Primary analysis from the Women's Health Initiative Observational Study and meta-analysis. *Cancer Epidemiol. Biomark. Prev.* **2011**, *20*, 1822–1830. [CrossRef] [PubMed]

18. Steinbrenner, H.; Speckmann, B.; Sies, H. Toward understanding success and failures in the use of selenium for cancer prevention. *Antioxid. Redox Signal.* **2013**, *19*, 181–191. [CrossRef] [PubMed]

19. Meplan, C.; Hughes, D.J.; Pardini, B.; Naccarati, A.; Soucek, P.; Vodickova, L.; Hlavata, I.; Vrana, D.; Vodicka, P.; Hesketh, J.E.; et al. Genetic variants in selenoprotein genes increase risk of colorectal cancer. *Carcinogenesis* **2010**, *31*, 1074–1079. [CrossRef] [PubMed]

20. Sutherland, A.; Kim, D.H.; Relton, C.; Ahn, Y.O.; Hesketh, J. Polymorphisms in the selenoprotein S and 15-kDa selenoprotein genes are associated with altered susceptibility to colorectal cancer. *Genes Nutr.* **2010**, *5*, 215–223. [CrossRef]

21. Meplan, C. Selenium and chronic diseases: A nutritional genomics perspective. *Nutrients* **2015**, *7*, 3621–3651. [CrossRef]

22. Slattery, M.L.; Lundgreen, A.; Welbourn, B.; Corcoran, C.; Wolff, R.K. Genetic variation in selenoprotein genes, lifestyle, and risk of colon and rectal cancer. *PLoS ONE* **2012**, *7*, e37312. [CrossRef]

23. Haug, U.; Poole, E.M.; Xiao, L.; Curtin, K.; Duggan, D.; Hsu, L.; Makar, K.W.; Peters, U.; Kulmacz, R.J.; Potter, J.D.; et al. Glutathione peroxidase tagSNPs: Associations with rectal cancer but not with colon cancer. *Genes Chromosomes Cancer* **2012**, *51*, 598–605. [CrossRef]

24. Riboli, E.; Kaaks, R. The EPIC Project: Rationale and study design. European Prospective Investigation into Cancer and Nutrition. *Int. J. Epidemiol.* **1997**, *26*, S6–S14. [CrossRef]

25. Riboli, E.; Hunt, K.J.; Slimani, N.; Ferrari, P.; Norat, T.; Fahey, M.; Charrondière, U.R.; Hémon, B.; Casagrande, C.; Vignat, J.; et al. European Prospective Investigation into Cancer and Nutrition (EPIC): Study populations and data collection. *Public Health Nutr.* **2002**, *5*, 1113–1124. [CrossRef]

26. Caboux, E.; Lallemand, C.; Ferro, G.; Hémon, B.; Mendy, M.; Biessy, C.; Sims, M.; Wareham, N.; Britten, A.; Boland, A.; et al. Sources of pre-analytical variations in yield of DNA extracted from blood samples: Analysis of 50,000 DNA samples in EPIC. *PLoS ONE* **2012**, *7*, e39821. [CrossRef] [PubMed]

27. Hutter, C.M.; Chang-Claude, J.; Slattery, M.L.; Pflugeisen, B.M.; Lin, Y.; Duggan, D.; Nan, H.; Lemire, M.; Rangrej, J.; Figueiredo, J.C.; et al. Characterization of gene-environment interactions for colorectal cancer susceptibility loci. *Cancer Res.* **2012**, *72*, 2036–2044. [CrossRef]

28. Meplan, C.; Rohrmann, S.; Steinbrecher, A.; Schomburg, L.; Jansen, E.; Linseisen, J.; Hesketh, J. Polymorphisms in thioredoxin reductase and selenoprotein K genes and selenium status modulate risk of prostate cancer. *PLoS ONE* **2012**, *7*, e48709. [CrossRef]

29. Hybsier, S.; Schulz, T.; Wu, Z.; Demuth, I.; Minich, W.B.; Renko, K.; Rijntjes, E.; Köhrle, J.; Strasburger, C.J.; Steinhagen-Thiessen, E.; et al. Sex-specific and inter-individual differences in biomarkers of selenium status identified by a calibrated ELISA for selenoprotein P. *Redox Biol.* **2017**, *11*, 403–414. [CrossRef]

30. Clarke, G.M.; Anderson, C.A.; Pettersson, F.H.; Cardon, L.R.; Morris, A.P.; Zondervan, K.T. Basic statistical analysis in genetic case-control studies. *Nat. Protoc.* **2011**, *6*, 121–133. [CrossRef]

31. Benjamini, Y.; Hochberg, Y. Controlling the false discovery rate: A practical and powerful approach to multiple testing. *J. R. Stat. Soc. B* **1995**, *57*, 289–300. [CrossRef]

32. Conneely, K.N.; Boehnke, M. So many correlated tests, so little time! Rapid adjustment of P values for multiple correlated tests. *Am. J. Hum. Genet.* **2007**, *81*, 1158–1168. [CrossRef] [PubMed]

33. Yu, K.; Li, Q.; Bergen, A.W.; Pfeiffer, R.M.; Rosenberg, P.S.; Caporaso, N.; Kraft, P.; Chatterjee, N. Pathway analysis by adaptive combination of P-values. *Genet. Epidemiol.* **2009**, *33*, 700–709. [CrossRef] [PubMed]

34. Zhang, H.; Shi, J.; Liang, F.; Wheeler, W.; Stolzenberg-Solomon, R.; Yu, K. A fast multilocus test with adaptive SNP selection for large-scale genetic-association studies. *Eur. J. Hum. Genet.* **2014**, *22*, 696–702. [CrossRef]

35. Dato, S.; De Rango, F. Antioxidants and Quality of Aging: Further Evidences for a Major Role of TXNRD1 Gene Variability on Physical Performance at Old Age. *Oxid. Med. Cell Longev.* **2015**, *2015*, 926067. [CrossRef]

36. Peters, U.; Hutter, C.M.; Hsu, L.; Schumacher, F.R.; Conti, D.V.; Carlson, C.S.; Edlund, C.K.; Haile, R.W.; Gallinger, S.; Zanke, B.W.; et al. Meta-analysis of new genome-wide association studies of colorectal cancer risk. *Hum. Genet.* **2012**, *131*, 217–234. [CrossRef]

37. Slattery, M.L.; Lundgreen, A.; Herrick, J.S.; Kadlubar, S.; Caan, B.J.; Potter, J.D.; Wolff, R.K. Genetic variation in bone morphogenetic protein and colon and rectal cancer. *Int. J. Cancer* **2012**, *130*, 653–664. [CrossRef] [PubMed]

38. Peters, U.; Bien, S.; Zubair, N. Genetic architecture of colorectal cancer. *Gut* **2015**, *64*, 1623–1636. [CrossRef] [PubMed]

39. Peters, K.M.; Carlson, B.A.; Gladyshev, V.N.; Tsuji, P.A. Selenoproteins in colon cancer. *Free Radic. Biol. Med.* **2018**, *127*, 14–25. [CrossRef]

40. Pitts, M.W.; Hoffmann, P.R. Endoplasmic reticulum-resident selenoproteins as regulators of calcium signaling and homeostasis. *Cell Calcium* **2018**, *70*, 76–86. [CrossRef]

41. Haggar, F.A.; Boushey, R.P. Colorectal cancer epidemiology: Incidence, mortality, survival, and risk factors. *Clin. Colon Rectal Surg.* **2009**, *22*, 191–197. [CrossRef] [PubMed]

42. Ekoue, D.N.; Zaichick, S.; Valyi-Nagy, K.; Picklo, M.; Lacher, C.; Hoskins, K.; Warso, M.A.; Bonini, M.G.; Diamond, A.M. Selenium levels in human breast carcinoma tissue are associated with a common polymorphism in the gene for SELENOP (Selenoprotein P). *J. Trace Elem. Med. Biol.* **2017**, *39*, 227–233. [CrossRef]

43. Evans, D.M.; Zhu, G.; Dy, V.; Heath, A.C.; Madden, P.A.; Kemp, J.P.; McMahon, G.; St Pourcain, B.; Timpson, N.J.; Golding, J.; et al. Genome-wide association study identifies loci affecting blood copper, selenium and zinc. *Hum. Mol. Genet.* **2013**, *22*, 3998–4006. [CrossRef]

44. Hazra, A.; Chanock, S.; Giovannucci, E.; Cox, D.G.; Niu, T.; Fuchs, C.; Willett, W.C.; Hunter, D.J. Large-scale evaluation of genetic variants in candidate genes for colorectal cancer risk in the Nurses' Health Study and the Health Professionals' Follow-up Study. *Cancer Epidemiol. Biomark. Prev.* **2008**, *17*, 311–319. [CrossRef] [PubMed]

45. Karunasinghe, N.; Han, D.Y.; Zhu, S.; Yu, J.; Lange, K.; Duan, H.; Medhora, R.; Singh, N.; Kan, J.; Alzaher, W.; et al. Serum selenium and single-nucleotide polymorphisms in genes for selenoproteins: Relationship to markers of oxidative stress in men from Auckland, New Zealand. *Genes Nutr.* **2012**, *7*, 179–190. [CrossRef] [PubMed]

46. Chiu-Ugalde, J.; Theilig, F.; Behrends, T.; Drebes, J.; Sieland, C.; Subbarayal, P.; Köhrle, J.; Hammes, A.; Schomburg, L.; Schweizer, U. Mutation of megalin leads to urinary loss of selenoprotein P and selenium deficiency in serum, liver, kidneys and brain. *Biochem. J.* **2010**, *431*, 103–111. [CrossRef] [PubMed]

47. Krol, M.B.; Gromadzinska, J.; Wasowicz, W. SeP, ApoER2 and megalin as necessary factors to maintain Se homeostasis in mammals. *J. Trace Elem. Med. Biol.* **2012**, *26*, 262–266. [CrossRef]

48. Vargas, T.; Bullido, M.J.; Martinez-Garcia, A.; Antequera, D.; Clarimon, J.; Rosich-Estrago, M.; Martin-Requero, A.; Mateo, I.; Rodriguez-Rodriguez, E.; Vilella-Cuadrada, E.; et al. A megalin polymorphism associated with promoter activity and Alzheimer's disease risk. *Am. J. Med. Genet. B Neuropsychiatr. Genet.* **2010**, *153B*, 895–902. [CrossRef]

49. Burk, R.F.; Hill, K.E.; Motley, A.K.; Winfrey, V.P.; Kurokawa, S.; Mitchell, S.L.; Zhang, W. Selenoprotein P and apolipoprotein E receptor-2 interact at the blood-brain barrier and also within the brain to maintain an essential selenium pool that protects against neurodegeneration. *FASEB J.* **2014**, *28*, 3579–3588. [CrossRef]

50. Franke, A.; McGovern, D.P.; Barrett, J.C.; Wang, K.; Radford-Smith, G.L.; Ahmad, T.; Lees, C.W.; Balschun, T.; Lee, J.; Roberts, R.; et al. Genome-wide meta-analysis increases to 71 the number of confirmed Crohn's disease susceptibility loci. *Nat. Genet.* **2010**, *42*, 1118–1125. [CrossRef]

51. Jostins, L.; Ripke, S.; Weersma, R.K.; Duerr, R.H.; McGovern, D.P.; Hui, K.Y.; Lee, J.C.; Schumm, L.P.; Sharma, Y.; Anderson, C.A.; et al. Host-microbe interactions have shaped the genetic architecture of inflammatory bowel disease. *Nature* **2012**, *491*, 119–124. [CrossRef] [PubMed]

52. Gentschew, L.; Bishop, K.S.; Han, D.Y.; Morgan, A.R.; Fraser, A.G.; Lam, W.J.; Karunasinghe, N.; Campbell, B.; Ferguson, L.R. Selenium, selenoprotein genes and Crohn's disease in a case-control population from Auckland, New Zealand. *Nutrients* **2012**, *4*, 1247–1259. [CrossRef] [PubMed]

53. Kipp, A.; Banning, A.; van Schothorst, E.M.; Méplan, C.; Schomburg, L.; Evelo, C.; Coort, S.; Gaj, S.; Keijer, J.; Hesketh, J.; et al. Four selenoproteins, protein biosynthesis, and Wnt signalling are particularly sensitive to limited selenium intake in mouse colon. *Mol. Nutr. Food Res.* **2009**, *53*, 1561–1572. [CrossRef] [PubMed]

54. Meplan, C.; Johnson, I.T.; Polley, A.C.; Cockell, S.; Bradburn, D.M.; Commane, D.M.; Arasaradnam, R.P.; Mulholland, F.; Zupanic, A.; Mathers, J.C.; et al. Transcriptomics and proteomics show that selenium affects inflammation, cytoskeleton, and cancer pathways in human rectal biopsies. *FASEB J.* **2016**, *30*, 2812–2825. [CrossRef] [PubMed]

55. Sun, R.; Jia, F.; Liang, Y.; Li, L.; Bai, P.; Yuan, F.; Gao, L.; Zhang, L. Interaction analysis of IL-12A and IL-12B polymorphisms with the risk of colorectal cancer. *Tumour Biol.* **2015**, *36*, 9295–9301. [CrossRef]

56. Sheng, W.Y.; Yong, Z.; Yun, Z.; Hong, H.; Hai, L.L. Toll-like receptor 4 gene polymorphisms and susceptibility to colorectal cancer: A meta-analysis and review. *Arch. Med. Sci.* **2015**, *11*, 699–707. [CrossRef] [PubMed]

57. Wang, P.; An, J.; Zhu, Y.; Wan, X.; Zhang, H.; Xi, S.; Li, S. Association of three promoter polymorphisms in interleukin-10 gene with cancer susceptibility in the Chinese population: A meta-analysis. *Oncotarget* **2017**, *8*, 62382–62399. [CrossRef] [PubMed]

58. Tsilidis, K.K.; Helzlsouer, K.J.; Smith, M.W.; Grinberg, V.; Hoffman-Bolton, J.; Clipp, S.L.; Visvanathan, K.; Platz, E.A. Association of common polymorphisms in IL10, and in other genes related to inflammatory response and obesity with colorectal cancer. *Cancer Causes Control* **2009**, *20*, 1739–1751. [CrossRef] [PubMed]

59. Miao, H.K.; Chen, L.P.; Cai, D.P.; Kong, W.J.; Xiao, L.; Lin, J. MSH3 rs26279 polymorphism increases cancer risk: A meta-analysis. *Int. J. Clin. Exp. Pathol.* **2015**, *8*, 11060–11067. [PubMed]

60. Naccarati, A.; Pardini, B.; Hemminki, K.; Vodicka, P. Sporadic colorectal cancer and individual susceptibility: A review of the association studies investigating the role of DNA repair genetic polymorphisms. *Mutat. Res.* **2007**, *635*, 118–145. [CrossRef]

61. Zhang, B.; Jia, W.H.; Matsuda, K. Large-scale genetic study in East Asians identifies six new loci associated with colorectal cancer risk. *Nat. Genet.* **2014**, *46*, 533–542. [CrossRef] [PubMed]

62. Wang, Y.; Yang, H.; Li, L.; Xia, X. An updated meta-analysis on the association of TGF-beta1 gene promoter -509C/T polymorphism with colorectal cancer risk. *Cytokine* **2013**, *61*, 181–187. [CrossRef] [PubMed]

63. Shanmugam, K.S.; Brenner, H.; Hoffmeister, M.; Chang-Claude, J.; Burwinkel, B. The functional genetic variant Arg324Gly of frizzled-related protein is associated with colorectal cancer risk. *Carcinogenesis* **2007**, *28*, 1914–1917. [CrossRef] [PubMed]

64. Berndt, S.I.; Huang, W.Y.; Yeager, M.; Weissfeld, J.L.; Chanock, S.J.; Hayes, R.B. Genetic variants in frizzled-related protein (FRZB) and the risk of colorectal neoplasia. *Cancer Causes Control* **2009**, *20*, 487–490. [CrossRef]

65. Hong, Y.; Wu, G.; Li, W.; Liu, D.; He, K. A comprehensive meta-analysis of genetic associations between five key SNPs and colorectal cancer risk. *Oncotarget* **2016**, *7*, 73945–73959. [CrossRef]

66. Lippman, S.M.; Klein, E.A.; Goodman, P.J.; Lucia, M.S.; Thompson, I.M.; Ford, L.G.; Parnes, H.L.; Minasian, L.M.; Gaziano, J.M.; Hartline, J.A.; et al. Effect of selenium and vitamin E on risk of prostate cancer and other cancers: The Selenium and Vitamin E Cancer Prevention Trial (SELECT). *JAMA* **2009**, *301*, 39–51. [CrossRef] [PubMed]

Article

Expression of Selenoprotein Genes and Association with Selenium Status in Colorectal Adenoma and Colorectal Cancer

David J. Hughes [1,*], Tereza Kunická [2], Lutz Schomburg [3], Václav Liška [2,4], Niall Swan [5] and Pavel Souček [2,4]

[1] Cancer Biology and Therapeutics Group, UCD Conway Institute, University College Dublin, D04 V1W8 Dublin, Ireland

[2] Biomedical Centre, Medical and Teaching School Pilsen, Charles University in Prague, 323 00 Pilsen, Czech Republic; terez.kunicka@gmail.com (T.K.); vena.liska@skaut.cz (V.L.); pavel.soucek@szu.cz (P.S.)

[3] Institute for Experimental Endocrinology, University Medical School Berlin, D-13353 Berlin, Germany; lutz.schomburg@charite.de

[4] Teaching Hospital and Medical School, Charles University in Prague, 306 05 Pilsen, Czech Republic

[5] Department of Pathology and Laboratory Medicine, St. Vincent's University Hospital, D04 T6F4 Dublin, Ireland; n.swan@svuh.ie

* Correspondence: david.hughes@ucd.ie; Tel.: +353-1-716-6988

Received: 10 September 2018; Accepted: 14 November 2018; Published: 21 November 2018

Abstract: Dietary selenium (Se) intake is essential for synthesizing selenoproteins that are important in countering oxidative and inflammatory processes linked to colorectal carcinogenesis. However, there is limited knowledge on the selenoprotein expression in colorectal adenoma (CRA) and colorectal cancer (CRC) patients, or the interaction with Se status levels. We studied the expression of seventeen Se pathway genes (including fifteen of the twenty-five human selenoproteins) in RNA extracted from disease-normal colorectal tissue pairs, in the discovery phase of sixty-two CRA/CRC patients from Ireland and a validation cohort of a hundred and five CRC patients from the Czech Republic. Differences in transcript levels between the disease and paired control mucosa were assessed by the Mann-Whitney U-test. *GPX2* and *TXNRD3* showed a higher expression and *GPX3*, *SELENOP*, *SELENOS*, and *SEPHS2* exhibited a lower expression in the disease tissue from adenomas and both cancer groups (*p*-values from 0.023 to <0.001). In the Czech cohort, up-regulation of *GPX1*, *SELENOH*, and *SOD2* and down-regulation of *SELENBP1*, *SELENON*, and *SELENOK* (*p*-values 0.036 to <0.001) was also observed. We further examined the correlation of gene expression with serum Se status (assessed by Se and selenoprotein P, SELENOP) in the Irish patients. While there were no significant correlations with both Se status markers, *SELENOF*, *SELENOK*, and *TXNRD1* tumor tissue expression positively correlated with Se, while *TXNRD2* and *TXNRD3* negatively correlated with *SELENOP*. In an analysis restricted to the larger Czech CRC patient cohort, Cox regression showed no major association of transcript levels with patient survival, except for an association of higher *SELENOF* gene expression with both a lower disease-free and overall survival. Several selenoproteins were differentially expressed in the disease tissue compared to the normal tissue of both CRA and CRC patients. Altered selenoprotein expression may serve as a marker of functional Se status and colorectal adenoma to cancer progression.

Keywords: selenium (Se), selenoproteins; gene expression; selenium status; selenoprotein P; colorectal neoplasm; colorectal cancer; colorectal adenoma; biomarkers; cancer risk

1. Introduction

Worldwide, colorectal cancer (CRC) is the second most common cancer in women and the third in men, with highest prevalence in developed countries [1]. Modifiable dietary and lifestyle patterns are important contributors to CRC etiology [2].

In humans, the essential micronutrient selenium (Se) is incorporated as the amino acid selenocysteine in twenty-five selenoproteins, of which several are involved in a wide variety of metabolic pathways implicated in carcinogenesis [3–5]. Observational and intervention studies suggest that Se status can influence colorectal adenoma (CRA) and colorectal cancer (CRC) development risk, particularly in the geographical areas of suboptimal Se availability, such as much of Europe [6,7]. Transcriptomic and proteomic studies demonstrate that Se intake can influence the pattern of selenoprotein expression and biosynthesis, affecting numerous oxidative stress, inflammatory, and signal translation pathways that are important in colorectal carcinogenesis [8–10]. Due to the hierarchical pattern of organ-specific selenoprotein expression in conditions of limited Se supply [3,11], tumor tissue specific expression patterns may provide a biologically informative marker of CRC risk, especially in relation to inadequate Se status [8,12].

However, there is limited data on the selenoprotein expression in CRA and CRC patient tissue, or the relationship with Se status. Human studies have generally only examined gene or protein colonic carcinoma tissue expression of key antioxidant selenoprotein P (SELENOP), glutathione peroxidase (GPX), and thioredoxin reductase (TXNRD) selenoenzymes that have different hierarchical expression regulation depending on dietary Se [3,8]. These studies have generally shown overlapping patterns of expression, with some differences in areas of lower [13,14] to higher [15] Se availability.

To examine the pattern of selenoprotein gene expression in CRA and CRC disease tissue and the surrounding non-neoplastic mucosa, the mRNA transcript expression level of seventeen Se pathway genes (including fifteen selenoproteins, the related *SELENBP1* Se biosynthesis gene, and the interacting *SOD2* antioxidant gene [16]) was evaluated for one hundred and sixty-seven Irish and Czech individuals (forty CRA and a hundred and twenty-seven CRC patients) by quantitative real-time PCR (qPCR). We were further able to compare the gene expression levels with two serum Se status markers (Se and selenoprotein P; SELENOP) in the Irish cohort. Potentially, selenoprotein mRNA level may be an easily measured molecular biomarker for assessing the biologically relevant Se status [17] for protection of healthy colonic tissue.

2. Materials and Methods

2.1. Patient Characteristics

This project consisted of a discovery and validation phase. In the discovery set of samples (Irish cohort), paired tumor tissue and adjacent non-neoplastic mucosa samples (and matched sera, where available) were obtained from a total of sixty-two patients with colorectal carcinoma (n = 22) or adenoma (n = 40), diagnosed at the Departments of Gastroenterology and Surgery, The Adelaide and Meath Hospital, in Dublin. The validation set (Czech cohort) comprised a hundred and five colorectal carcinoma patients, diagnosed at the Department of Surgery, Teaching Hospital and Medical School, in Pilsen, between January 2008 and November 2011. The clinical characteristics of our study cohorts are summarized in Table 1.

Irish tissue samples were collected during resection of primary tumor or by biopsy, before treatment, while adenoma biopsies were obtained at colonoscopy during a pilot CRC screening program as described previously [18]. Only advanced CRAs were included in the study—forty tubular or tubulovillous adenomas of at least 1cm, of which fourteen had high grade dysplasia (HGD), as the most screen-relevant lesions [19]. Blood samples were collected immediately within one day of surgery or colonoscopy, in plain 6 mL VACUTAINER® tubes, (Cruinn Diagnostics, Dublin, Ireland) containing no anticoagulant. Blood was centrifuged at 2000× *g* for 10 min, within 4 h of collection, to separate the serum layer, which was then stored at −80 °C in cryovials. Collection and pathological processing of tissue samples and retrieval of data acquisition of the Czech samples, was performed as previously described [20]. Histology was verified by an experienced pathologist at each center. All CRCs were classified according to the tenth revision of the International Classification of Diseases (ICD-10) and the second revision of the International Classification of Disease for Oncology (ICDO-2). The clinical data, including age at diagnosis, sex, pTNM (Tumor stage, Regional lymph node involvement, and distant metastasis) staging, histological grade of the tumor, and primary tumor localization were taken from patient medical records (see Table 1).

Table 1. Clinical-pathological characteristics of the studied cohort of patients.

Cohort	Irish	Irish	Czech
Diagnosis	CRA	CRC	CRC
Total n of tissue samples	40	22	105
Sex n (male/female)	26/14	11/11	64/25
Age at diagnosis, median ± SD (years)	61 ± 7 years	59 ± 11 years	69 ± 11 years
Location n (colon/rectum)	28/12	15/7	82/17
T staging n (/T1/T2/T3/T4)	-	6/4/9/3	-
N staging n (N0/N1 or N2) (missing)	-	19/3	78/25 (2)
M staging n (M0/M1)	-	21/1	-
Stage (I/II/III/IV (missing))	-	10/7/4/1	2/76/15/10 (2)

SD = standard deviation; CRC = colorectal cancer; CRA = colorectal adenoma; - = not applicable or missing; TNM staging = Tumor stage, Regional lymph node involvement, and distant metastasis.

All patients were asked to carefully read and sign an informed consent, in accord with the 1964 Helsinki Declaration. The study was approved by the Ethical Committee of the St. James's Hospital and Federated Dublin Voluntary Hospitals Joint Research Ethics Committee (Ireland, reference 2007-37-17), and the Ethical Committee of the Medical Faculty and Teaching Hospital in Pilsen (Czech Republic, reference NT12025-4/2011). All samples were coded to protect patient anonymity.

2.2. Isolation of Total RNA

In the Irish cohort, the tissue samples were lysed on ice in 400 uL of lysis buffer (50 mmol/L HEPES, 4-(2-hydroxyethyl)-1-piperazineethanesulfonic acid, pH 7.5; 150 mmol/L NaCl; 5 mmol/L EDTA) and protease inhibitor (Calbiochem, Hampshire, UK) followed by sonication (3 × 30 s, on ice) and centrifugation (10,000× g for 10 min at 4 °C). Total RNA (as well as gDNA and protein) was then extracted using the Norgen All-in-one purification kit (Norgen, Thorold, ON, Canada). In the Czech cohort, tissue samples were homogenized by a mechanical disruption in liquid nitrogen using mortar and pestle, and the total RNA was isolated using the Trizol Reagent according to the manufacturer protocol (Invitrogen, Carlsbad, CA, USA). RNA was then stored at minus 80 °C and the quantity and quality were measured, as previously described [21].

2.3. Synthesis of Complementary DNA (cDNA)

Reverse transcription of the total RNA (0.5 µg for each reaction) was performed, using random hexamer primers and the RevertAid™ First Strand cDNA Synthesis Kit (MBI Fermentas, Vilnius, Lithuania). Quality of the cDNA, in terms of DNA contamination, was confirmed by PCR amplification of *Ubiquitin C* [22].

2.4. Relative Quantification of Gene Expression

qPCR was performed using ViiA7 Real-Time PCR System, using TaqMan® Universal Master Mix and TaqMan® Gene Expression Assays (Life Technologies, Carlsbad, CA, USA) optimized primer and probe sets for the fifteen selenoprotein genes *SELENOP, SELENOS, GPX1-4, SELENOF, SEPHS2, SELENOH, SELENOK, SELENON, SELENOW, TXNRD1-3*, plus the Se biosynthesis gene *SELENBP1* and the interacting *SOD2* gene, in oxidative defense (Supplementary Table S1).

POLR2A, PSMC4, and MRPL19 were used as the reference genes, based on stability assessment of twenty-four potential endogenous control genes in a test set of ten pairs of CRC tumors and non-neoplastic tissue samples [20]. Efficiency of the qPCR was determined for each assay using calibration curves (six-point, five times dilution), which were prepared from one non-neoplastic sample for the Czech cohort and from the mix of samples for the Irish cohort, respectively. The non-template control contained nuclease-free water instead of cDNA. Samples with variation larger than 0.5 Cq (quantitation cycle) were reanalyzed. The qPCR study adhered to the Minimum Information for Publication of Quantitative Real-Time PCR Experiments (MIQE) Guidelines [23].

2.5. Serum Selenium and Selenoprotein P Measurements

The sample type (cancer, adenoma, and control) was blinded. Concentrations of the total Se were measured in 4 µL of each serum sample, using a bench-top total reflection X-ray fluorescence (TXRF) spectrometer (PicofoxTM S2, Bruker Nano GmbH, Berlin, Germany), as described previously [7]. SELENOP levels were ascertained by a colorimetric enzyme-linked immunoassay (SelenotestTM, ICI GmbH, Berlin, Germany) using 5 µL of each serum sample in a 1:21 dilution, according to the manufacturer's instructions [24]. Duplicate samples with differences in the Se concentration varying by more than 10% were measured again. The evaluation was performed with the GraphPad Prism 6.01 (GraphPad Software, La Jolla, CA, USA), using a four-parameter logistic function. The samples were measured in duplicate, and the mean concentration values, standard deviation (SD), and coefficients of variation (CVs) were calculated. The CVs were below 10% for the *SELENOP* controls 1 (1.5 mg/L) and 2 (8.6 mg/L) throughout the measurements.

2.6. Statistical Analyses

Transcript levels were analyzed by Viia7 System Software (Life Technologies, Carlsbad, CA, USA) and normalized levels of target genes (ratio of target gene Cq to the mean Cq of reference genes) were used for further statistical analyses, using SPSS v16.0 Software (SPSS Inc., Chicago, IL, USA). Differences in transcript levels between tumor and control tissues were assessed by nonparametric Mann-Whitney U-test. Nonparametric tests (the Kruskal-Wallis, the Mann-Whitney, and the Spearman's test) were also used for evaluation of associations of transcript levels with clinical data and other variables (Table 1). Kaplan-Meier Log Rank test and multivariate Cox regression analyses, adjusted to stage, surgical radicality, and chemotherapy, were used to assess the gene transcript levels, with subject disease-free survival (DFS) and overall survival (OS), in the Czech CRC cohort. Cox regression was used to determine the hazard ratio (HR) and the 95% confidence intervals (95% CI) of the transcript levels, were divided according to the median, with these survival outcomes.

Analyses were conducted using SPSS v16.0 Software (SPSS Inc., Chicago, IL, USA). All *p*-values were obtained from two-sided tests; *p*-values lower than 0.05 were considered statistically significant. Multiple testing adjustment (P_{ADJ}) was performed by the Benjamini-Hochberg correction (BH).

3. Results

3.1. Selenoprotein Gene Transcript Levels in the Colorectal Adenomas

In the Irish cohort, CRA patients showed a strongly significant lower expression of *SELENOP*, *SELENOS*, *GPX3*, and *SEPHS2* and higher expression levels of *GPX2*, *SELENOH*, and *TXNRD3* ($p < 0.001$ for all genes), in the disease tissue compared to the non-neoplastic control tissues. We tested the trend for expression with adenoma progression, compared to the normal tissues, by splitting the CRAs into two groups of twenty-six tubular and/or villous adenomas and another of fourteen HGDs. This suggested a declining expression of *SELENOP*, *SELENOS*, *GPX3*, and *SEPHS2* (respective *p*-values for trend at least 5.1×10^{-5}) and increasing expression of *GPX2*, *SELENOH*, and *TXNRD3* (respective *p*-values at least 7×10^{-5}), from the control to the increasingly dysplastic tissues.

3.2. Selenoprotein Gene Transcript Levels in Colorectal Cancers

In the tumor tissues of the CRC patients from Ireland, compared to the matched controls, *GPX2*, *GPX4*, *TXNRD3*, and *SOD2* were up-regulated (p = 0.023, 0.039, 0.003, and 0.036, respectively), while *SELENOP*, *SELENOS*, *GPX3*, *SEPHS2*, *SELENBP1*, and *SELENOK* were considerably down-regulated (p = 0.001, <0.001, <0.001, 0.002, <0.001, and 0.015, respectively).

In the validation study (Czech cohort), while many of the above observations were replicated, several disparities were observed. The Irish patients showed overexpressed levels of *GPX4* in tumors, compared to the control tissues, while the Czech subjects had no difference. Up-regulation of *GPX1* and *SELENOH* (p = 0.001 for both genes), and down-regulation of *SELENON* (p = 0.001) were demonstrated in the Czech patients, but no significant changes in the expression levels of these genes were found in the Irish cohort (although *SELENOH* was significantly more highly expressed in the adenoma tissue from the Irish CRA patients, p < 0.001).

All statistically significant correlations from the three groups of patients are summarized in Table 2. Directions of expression (higher or lower) were fully consistent for significant changes in the same gene across the different patient groups from Ireland and the Czech Republic. Gene expression levels did not correlate with the RNA integrity number (RIN), suggesting that RNA quality did not significantly influence the results.

Table 2. Differences in the transcript levels of the examined genes between the colorectal adenoma/carcinoma and the non-neoplastic tissue in the Irish and the Czech CRA/CRC patients (this study), and comparison with data from the TCGA patient cohorts.

| Gene Name | Discovery (Irish Cohort, n = 62; 40 CRA and 22 CRC) | | Validation (Czech Cohort, N = 105 CRC) | TCGA (COAD/READ) |
	Adenoma Expression (p-Value)	Carcinoma Expression (p-Value)	Carcinoma Expression (p-Value)	Carcinoma Expression
GPX1	NS	NS	↑ (<0.001)	NS/NS
GPX2	↑ (<0.001)	↑ (0.023)	↑ (<0.001)	↑/↑
GPX3	↓ (<0.001)	↓ (<0.001)	↓ (<0.001)	↓/↓
GPX4	NS	↑ (0.039)	NS	NS/NS
SELENOF	NS	NS	NS	NS/NS
SELENOH	↑ (<0.001)	NS	↑ (<0.001)	NS/NS
SELENOK	NS	↓ (0.015)	↓ (<0.001)	NS/NS
SELENON	NS	NS	↓ (<0.001)	NS/NS
SELENOP	↓ (<0.001)	↓ (0.001)	↓ (<0.001)	↓/↓
SELENOS	↓ (<0.001)	↓ (<0.001)	↓ (<0.001)	NS/NS
SELENOW	NS	NS	NS	NS/NS
SEPHS2	↓ (<0.001)	↓ (0.002)	↓ (<0.001)	NS/NS
TXNRD1	NS	NS	NS	NS/NS
TXNRD2	NS	NS	NS	NS/NS
TXNRD3	↑ (<0.001)	↑ (0.003)	↑ (<0.001)	↑/↑
SELENBP1	NS	↓ (<0.001)	↓ (<0.001)	↑/↑
SOD2	NS	↑ (0.036)	↑ (<0.001)	NS/NS

Abbreviations: ↑ = higher expression in the neoplastic compared to the normal tissue, ↓ = lower expression in the neoplastic compared to the normal tissue, NS = not significant, TCGA = The Cancer Genome Atlas (https://cancergenome.nih.gov/), COAD = colon adenocarcinoma, and READ = rectum adenocarcinoma. For the seventeen genes tested in the Irish and the Czech studies, only *SELENOF*, *SELENOW*, *TXNRD1*, and *TXNRD2* were not significantly different in at least in one of the neoplasm groups. Genes significant in all three Irish and Czech sample sets (*GPX2*, *GPX3*, *SELENOP*, *SELENOS*, *SEPHS2*, and *TXNRD3*) are marked by a dark background. p-values are in brackets. TCGA data: Analysis done on the TCGA tumor-normal data, using the GEPIA (Gene Expression Profiling Interactive Analysis) tool (http://gepia.cancer-pku.cn/about.html; Tang et al. Nucleic Acids Res. 2017; 45(W1): W98–W102).

There were no major consistent differences in the expression patterns by age, sex, or tumor sub-site (colon/rectum) across the Irish and the Czech CRC samples. Stratified analyses—in the larger Czech CRC group—of expression by tumor stage, grade, and TNM status, showed no significant differences after the BH correction, except for an association of a higher cancer stage and presence of lymph node metastasis with decreased *SOD2* expression (P_{ADJ} = 0.004 and P_{ADJ} = 0.048, respectively). The presence of lymph node metastasis also significantly associated with decreased *GPX1* expression

(P_{ADJ} = 0.048). As our data indicated that increasing numbers of selenoprotein genes were dysregulated from the adenoma to tumor tissue and, thus, possibly to be involved in disease progression, we examined gene expression with cancer patient outcome. We performed analyses of tumor selenoprotein transcript levels, divided by the median, with the DFS and the OS, using 8–76 months patient follow-up data that were available for ninety-eight of the Czech CRCs (mean values of 58 and 68 months, respectively). Only a higher *SELENOF* expression was associated with both a poorer DFS and OS, in a Kaplan-Meier (see Figure 1) and multivariate-adjusted Cox regression analysis (HR: 2.41; 95% CI: 1.08, 5.41; p = 0.032 for DFS, and HR: 4.13; 95% CI: 1.10, 15.39; p = 0.035 for OS). However, this result, although biologically-plausible based on previous data of the oncogenic potential of *SELENOF* [3], should be cautiously considered, as the CIs were wide and the point estimates lost significance after multiple testing correction. Additionally, *SELENOF* did not show significant differences in overall expression levels between the neoplastic and the matched mucosal tissues.

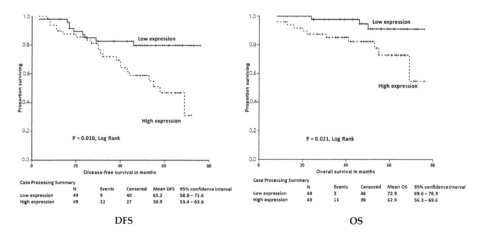

Figure 1. Association of the Selenoprotein F (*SELENOF*) gene expression with disease-free survival and overall survival, in colorectal cancer patients. Kaplan-Meier curves showing disease free survival (DFS) overall survival (OS) of 98 colorectal cancer (CRC) patients from the Czech Republic with higher *SELENOF* gene expression compared to low expression in tumor tissue. Subjects with higher *SELENOF* expression than median show poorer prognosis with shorter DFS and OS compared to subjects with lower *SELENOF* expression (50.9 vs. 65.2 months and 62.9 vs. 72.9 months, respectively; P_{DFS} = 0.01 and P_{OS} = 0.021). In multivariable Cox regression analyses adjusted by tumor stage, surgical radicality, and chemotherapy this was associated with a Hazards ratio (HR) point estimate of 2.41 for DFS (95% CI: 1.08, 5.41; P = 0.032) and 4.13 (95% CI: 1.10, 15.39; P = 0.035) for OS.

Most of the studied genes (12 of the 17) showed non-significant expression changes in The Cancer Genome Atlas (TCGA) dataset (https://cancergenome.nih.gov/) (see Supplementary Figure S1). The other five genes showed concordant significant expression changes across the Irish, Czech, and the TCGA datasets, including the main Se transport protein-coding gene *SELENOP*, *GPX2*, *GPX3* (the second major Se transport gene), *TNRDX3*, and *SELENBP1* (see last column in Table 2). As presently we only had tissue RNA available for analysis, we were not able to analyze the protein expression in our samples. However, Supplementary Table S2 presents a summary of the immunohistochemistry protein expression data from the Human Protein Atlas (HPA) dataset (https://www.proteinatlas.org/), for the corresponding seventeen genes examined in this study. Although the numbers of the tumor samples with the available data are too small (n = 9–12 for each selenoprotein) to make meaningful inferences, several proteins (GPX2, GPX3, TXNRD3) had expression changes in most samples that matched the direction of gene expression variance observed in both the Irish and the Czech CRC samples. *SELENOP*, *SELENBP1*, and *SOD2* had medium protein expression levels, in both the normal and the tumor cells. Contrary results were given only by *SELENOS* and *SEPHS2*, which appeared to have relatively higher protein expression levels in the tumor cells (although *SEPHS2* was not detected in the normal cells).

3.3. Correlation of the Selenoprotein Transcript Levels with Se Status

Spearman correlation coefficients were calculated to ascertain the correlation of the selenoprotein gene expression in the Irish cohort with two measures of serum Se status (Se and SELENOP); available for thirteen CRA and seventeen CRC (Se) and thirty-eight CRA and eighteen CRC (SELENOP) patients, respectively. Only positive correlations of gene expression with the Se level were observed for the *SELENOF*, *SELENOK*, and *TXNRD1* genes, all in the tumor tissue ($p = 0.001$, 0.004, and 0.04, respectively), except for the normal tissue expression of *SELENOK* as well ($p = 0.03$). Regarding the SELENOP level, *TXNRD2* and *TXNRD3* tumor tissue expression was negatively correlated ($p = 0.037$ and 0.045), while *GPX1* had a higher expression in the normal tissue, $p = 0.034$. The only significant findings for adenomas were negative correlations for the SELENOP levels with *TXNRD1* disease ($p = 0.006$) and *SELENOW* normal ($p = 0.042$) tissue expressions. All correlation values are available in Supplementary Table S3.

4. Discussion

Expression profiles of seventeen Se pathway genes (including fifteen selenoproteins) were assessed in adenoma and cancer tissues, with their respective matched normal tissues, for forty CRA and a hundred and twenty-seven CRC patients. In comparisons between the neoplastic and normal tissue pairs, we observed seven differentially expressed genes in the CRA patients from Ireland and twelve dysregulated Se pathway genes for the CRC patients from the Czech Republic. Genes up-regulated in the tumor tissues were *GPX1*, *GPX2*, *SELENOH*, *TXNRD3*, and *SOD2*, while those down-regulated included *GPX3*, *SELENOP*, *SELENOS*, *SEPHS2*, *SELENBP1*, *SELENON*, and *SELENOK*. In adenomas *GPX2*, *SELENOH*, and *TXNRD3* also exhibited a higher expression in the disease tissue, while *GPX3*, *SELENOP*, *SELENOS*, and *SEPHS2* showed a lower expression for the Czech cancers. Thus, *GPX1*, *SELENBP1*, *SELENON*, *SELENOK*, and *SOD2* were differently expressed only in the cancer tissues.

Broadly similar expression patterns were observed for several selenoprotein genes (e.g., *GPX2*, *GPX3*, *SELENOP*, *SELENBP1*, and *TNRDX3*) across our Irish and Czech study groups, as compared with the gene and protein data for the TCGA and HPA, respectively. There were non-significant expression changes in the other selenoprotein genes in the TCGA, compared to significant findings in all our sample sets (*SEPHS2*, *SELENOS*), although the trend for a lower expression in tumor tissue for these genes was also seen in the TCGA. However, this is discordant with the corresponding SEPHS2 and SELENOS protein data in the HPA. These findings are probably largely explained by differences in the baseline Se status of the heterogenous populations in the TCGA and HPA data, resulting in differential regulation within the Se hierarchy [8], the method used for gene expression analysis (the more insensitive RefSeq, compared to the qPCR, for subtle changes), and variances in expression regulation at the gene or protein level.

Selenoproteins have several well-demonstrated or suggested cell protection roles in pathways implicated in colorectal carcinogenesis, such as the antioxidant response, immune, and inflammatory pathways [25–27], while downstream-targeted metabolic pathways are also affected by Se status, as demonstrated in human rectal biopsies [9]. Thus, altered selenoprotein expression in the colorectal tract may also be affected by Se status could increase cancer development risk by weakening the gut epithelial cell response to harmful oxidative and inflammatory challenges [8]. Transcriptomics animal studies highlight *Gpx1*, *Selenof*, *Selenoh*, and *Selenow* as being sensitive to Se supply and human Single Nucleotide Polymorphism (SNP) studies suggest *SELENOP*, *SELENOS*, *GPX4*, *SELENOF*, *SELENON*, *SELENOH*, and *TXNRD1-3* are key selenoproteins for colonic function and colorectal carcinogenesis [8,16,27,28]. However, prior to our investigation, data were lacking on the expression of all these selenoproteins in CRA and CRC. In this study, changes in gene expression were observed for *SELENOP*, *SELENOS*, and *TXNRD3* for all sample groups, while *GPX4* was significant in the Irish cancers only and *GPX1*, *SELENOH*, and *SELENON* were differently regulated in the Czech CRCs. Additional genes in this study, showing changes in gene expression for both the Irish and the Czech cancers were *GPX2*, *GPX3*, *SELENOK*, *SEPHS2*, *SELENBP1*, and *SOD2*.

Only a higher expression of the *SELENOF* gene was associated with patient outcomes from cancer (lower DFS and OS), although this was non-significant after multiple testing adjustment. However, several other lines of evidence suggest that *SELENOF* may act as an Se-dependent oncogenic protein. *SELENOF* can be regulated by both Se status (as indicated also in our study) and endoplasmic reticulum (ER) stress, and several *SELENOF* genetic variants have been associated with an altered risk at the different cancer sites, including CRC [8,29]. Notably, previous cell and mouse-model studies suggest that higher levels of *SELENOF* may potentially contribute to favorable growth conditions for the cancerous cells in the colorectum, by reducing cellular ER stress and nuclear factor kappa-light-chain-enhancer of activated B cells (NFkB) activation [29–31].

Normally, selenoprotein expression is sensitive to limited Se supply in a tissue-dependent manner [11]. As expected in conditions of redox stress, a lowered *SELENOP* expression was reported in studies of a small number of German CRA and CRC subjects, while heterogenous *GPX2* mRNA and protein expression was observed between the tumor samples [13,32]. In German patients, *GPX1*, *GPX4*, and *TXNRD1* were also found to be up-regulated in cancer, compared to the matched tissues for mRNA and/or their corresponding protein levels [14]. However, while *GPX1* and *GPX4* were more highly expressed in tumor tissue from the Czech and Irish cancer groups, respectively, *TXNRD1* was not significantly different in any of our tested cohorts. In other settings of generally higher Se availability, such as in Japan, the protein expression of GPX1, GPX3, and SELENOP was reported to be lower in the CRC tumors, whereas the less Se-sensitive GPX2 was increased [15]. We observed similar results for our mRNA assays for *GPX2*, *GPX3*, and *SELENOP* for all CRA and CRC groups.

Hypoxic and oxidative stresses in proliferating tumors may also decouple the normal hierarchy of selenoprotein expression [12]. *SELENOH* is a putative redox regulating DNA-binding protein, whose cellular expression is sensitive to Se supply [26,33]. In contrast, protein expression of SELENOH has recently been shown to be higher in CRC human tumors (and to control cell-cycle progression and tumor proliferation in mouse and human CRC cell-line models) [12], aligning with the upregulation of *SELENOH* in the CRAs and the Czech CRCs observed in this study. The Se binding protein 1 (*SELENBP1*) gene and corresponding protein expression was found to be downregulated in Chinese CRC samples, as observed for the *SELENBP1* mRNA levels in the cancer cohorts in this study [34].

Overall, the apparent expression pattern we assessed in the colorectal tumor tissue is that the genes related to Se homeostasis (*SELENOP*, *SEPHS2*, *SELENBP1*) and ER stress (*SELENOK*, *SELENOS*) are down-regulated, while the antioxidant enzymes might exhibit a higher (*GPX1*, *GPX2*, *SELENOH*, *TXNRD3*, *SOD2*) or lower (*GPX3*, *SELENON*) expression [3,6,8]. In a previous examination of the common genetic variation in selenoproteins and Se-pathway genes, we observed that several SNPs in many of these genes were associated with CRC risk [35]. These data all support a role of selenoprotein metabolism and endoplasmic and oxidative stress in CRC development. Differences and overlaps between studies likely reflect design issues of sample size and tissue stage and site sampled, or underlying biological differences in the Se metabolism. This could be due to population specific differential regulation within the Se hierarchy, depending on the Se availability and selenoprotein genetic variation [8].

Several selenoprotein genes showed sequential expression differences through the major stages of dysplastic adenoma progression observed in different individuals in the Irish study. These included increasing down-regulation of *GPX3*, *SELENOP*, *SELENOS*, and *SEPHS2* and up-regulation of *GPX2*, *SELENOH*, and *TXNRD3*, in groups of normal-matched tissues, adenomas, and adenomas with HGD. Additional and larger studies will help clarify the pattern of selenoprotein expression in the adenoma progression. Possibly, selenoprotein expression may have uses as markers of advanced adenoma stages, relevant for CRC screening, especially as only advanced adenomas appear to be associated with an increased risk for subsequent CRC development [36].

Suboptimal Se status levels are found in many parts of Europe [37,38]. Our measures of Se status were only possible for available matching serum samples for the Irish CRA and CRC patients. Here, the mean levels of Se and *SELENOP* (86.1 µg/L in 30 sera and 5.1 mg/L in 56 sera, respectively) were slightly higher, as compared to eight other Western European countries in a prospective study, showing an association of higher levels with a decreased CRC risk (85.6 µg/L and 4.4 mg/L, respectively for the controls [7]). Although, there were no marked correlation patterns between either of these serum Se status levels and the selenoprotein gene expression, they did indicate the potential importance of the Se supply for expression of *GPX1*, *SELENOF*, *SELENOK*, and *TXNRD1-3*. Most of the few significant correlations were observed for the tumor tissue gene expression, rather than adenoma tissue or the matched normal tissues for these pathologies, suggesting that Se status is a factor for the differential expression we observed for *SELENOK* and *TXNRD3* in the Irish CRC group. Although we did not observe significant correlations of their expression with the Se status, other selenoprotein genes whose activities are thought to be sensitive to lower Se status in the gastrointestinal tract, such as *SELENOP* and *SELENOH* [8,12], were also observed to be differentially regulated in the adenomas and/or cancers groups from Ireland. However, the total numbers available for these analyses were small and confined to the Irish samples. Therefore, this should be considered as an exploratory analysis requiring further investigation to more fully ascertain the degree to which any involvement of selenoprotein expression in affecting CRA to CRC development may depend on the Se status of the studied population.

In summary, several selenoprotein genes were differentially expressed in both the CRA and CRC disease-normal tissue pairs. These include key biological stress response genes like *GPXs*, *TXNRD3*, *SELENOS*, *SELENOH*, and the related *SOD2* gene implicated in cancer cell survival, and genes such as *SELENOP* and *SEPHS2* involved in Se biosynthesis. However, as only some of these expression changes were correlated with the Se status levels, selenoprotein expression may affect CRA to CRC development independent of the Se status. Functional studies will be required to further investigate any role of the identified differentially-expressed selenoproteins in colorectal carcinogenesis. Potentially, selenoprotein mRNA expression may have uses as biomarkers of colorectal function, colorectal neoplasia progression, and improved assessment of physiological Se status, with implications for the modulation of Se intake.

Supplementary Materials: The following are available online at http://www.mdpi.com/2072-6643/10/11/1812/s1, Figure S1: Selenoprotein gene expression profiles from tumor-normal data of human colorectal tumor (red boxes) and mucosa (grey boxes) tissues from the TCGA database, Table S1: List of TaqMan selenoprotein and selenium-related gene expression assays used for this study, Table S2: Protein levels of selenoproteins in normal and colorectal cancer tissues in The Human Protein Atlas, Table S3: Spearman correlation values for tissue selenoprotein gene expression and serum selenium status in cancer and adenoma patients from Ireland.

Author Contributions: Conceptualization, D.J.H. and P.S.; Data curation, T.K., L.S., V.L. and N.S.; Formal analysis, D.J.H., T.K., L.S. and P.S.; Funding acquisition, D.J.H. and P.S.; Investigation, D.J.H., T.K., L.S. and P.S.; Methodology, D.J.H., T.K., L.S., V.L. and P.S.; Project administration, D.J.H., V.L., N.S. and P.S.; Resources, V.L., N.S. and P.S.; Software, T.K., L.S. and P.S.; Supervision, D.J.H. and P.S.; Validation, D.J.H. and L.S.; Writing–original draft, D.J.H. and T.K.; Writing–review & editing, L.S., N.S. and P.S.

Funding: This research was funded by project grants HRA/PHS/2013/397 and HRA/PHS/2015/1142 (Health Research Board of Ireland, P.I., D.J.H.), P303/12/G163 (the Czech Science Foundation to P.S.), LO1503 (National Sustainability Program I of the Ministry of Education Youth and Sports of the Czech Republic to T.K.), UNCE/MED/006 (Charles University project "Center of clinical and experimental liver surgery" to V.L.), and FNPl 00669806 (Conceptual Development of Research Organization-Faculty Hospital in Pilsen to V.L. and P.S.). L.S. was supported by Deutsche Forschungsgemeinschaft (DFG Research Unit 2558 TraceAge, Scho 849/6-1).

Conflicts of Interest: The authors declare no conflict of interest. While L.S. is the founder of selenOmed GmbH, a company involved in improving Se diagnostics, the data were analyzed before the founding of selenOmed and the analyses were conducted completely blinded to any clinical finding. The funders had no role in the design of the study; in the collection, analyses, or interpretation of data; in the writing of the manuscript, and in the decision to publish the results.

References

1. Ferlay, J.; Soerjomataram, I.; Dikshit, R.; Eser, S.; Mathers, C.; Rebelo, M.; Parkin, D.M.; Forman, D.; Bray, F. Cancer incidence and mortality worldwide: Sources, methods and major patterns in globocan 2012. *Int. J. Cancer* **2015**, *136*, E359–E386. [CrossRef] [PubMed]

2. Cappellani, A.; Zanghi, A.; Di Vita, M.; Cavallaro, A.; Piccolo, G.; Veroux, P.; Lo Menzo, E.; Cavallaro, V.; de Paoli, P.; Veroux, M.; et al. Strong correlation between diet and development of colorectal cancer. *Front. Biosci.* **2013**, *18*, 190–198.

3. Labunskyy, V.M.; Hatfield, D.L.; Gladyshev, V.N. Selenoproteins: Molecular pathways and physiological roles. *Physiol. Rev.* **2014**, *94*, 739–777. [CrossRef] [PubMed]

4. Hatfield, D.L.; Tsuji, P.A.; Carlson, B.A.; Gladyshev, V.N. Selenium and selenocysteine: Roles in cancer, health, and development. *Trends Biochem. Sci.* **2014**, *39*, 112–120. [CrossRef] [PubMed]

5. Steinbrenner, H.; Speckmann, B.; Klotz, L.O. Selenoproteins: Antioxidant selenoenzymes and beyond. *Arch. Biochem. Biophys.* **2016**, *595*, 113–119. [CrossRef] [PubMed]

6. Meplan, C. Selenium and chronic diseases: A nutritional genomics perspective. *Nutrients* **2015**, *7*, 3621–3651. [CrossRef] [PubMed]

7. Hughes, D.J.; Fedirko, V.; Jenab, M.; Schomburg, L.; Meplan, C.; Freisling, H.; Bueno-de-Mesquita, H.B.; Hybsier, S.; Becker, N.P.; Czuban, M.; et al. Selenium status is associated with colorectal cancer risk in the European prospective investigation of cancer and nutrition cohort. *Int. J. Cancer* **2015**, *136*, 1149–1161. [CrossRef] [PubMed]

8. Meplan, C.; Hesketh, J. Selenium and cancer: A story that should not be forgotten-insights from genomics. *Adv. Nutr. Cancer* **2014**, *159*, 145–166.

9. Meplan, C.; Johnson, I.T.; Polley, A.C.; Cockell, S.; Bradburn, D.M.; Commane, D.M.; Arasaradnam, R.P.; Mulholland, F.; Zupanic, A.; Mathers, J.C.; et al. Transcriptomics and proteomics show that selenium affects inflammation, cytoskeleton, and cancer pathways in human rectal biopsies. *FASEB J.* **2016**, *30*, 2812–2825. [CrossRef] [PubMed]

10. Peters, K.M.; Carlson, B.A.; Gladyshev, V.N.; Tsuji, P.A. Selenoproteins in colon cancer. *Free Radic. Biol. Med.* **2018**. [CrossRef] [PubMed]

11. Schomburg, L.; Schweizer, U. Hierarchical regulation of selenoprotein expression and sex-specific effects of selenium. *Biochim. Biophys. Acta* **2009**, *1790*, 1453–1462. [CrossRef] [PubMed]

12. Bertz, M.; Kuhn, K.; Koeberle, S.C.; Muller, M.F.; Hoelzer, D.; Thies, K.; Deubel, S.; Thierbach, R.; Kipp, A.P. Selenoprotein H controls cell cycle progression and proliferation of human colorectal cancer cells. *Free Radic. Biol. Med.* **2018**. [CrossRef] [PubMed]

13. Al-Taie, O.H.; Uceyler, N.; Eubner, U.; Jakob, F.; Mork, H.; Scheurlen, M.; Brigelius-Flohe, R.; Schottker, K.; Abel, J.; Thalheimer, A.; et al. Expression profiling and genetic alterations of the selenoproteins GI-GPx and SePP in colorectal carcinogenesis. *Nutr. Cancer* **2004**, *48*, 6–14. [CrossRef] [PubMed]

14. Yagublu, V.; Arthur, J.R.; Babayeva, S.N.; Nicol, F.; Post, S.; Keese, M. Expression of selenium-containing proteins in human colon carcinoma tissue. *Anticancer Res.* **2011**, *31*, 2693–2698. [PubMed]

15. Murawaki, Y.; Tsuchiya, H.; Kanbe, T.; Harada, K.; Yashima, K.; Nozaka, K.; Tanida, O.; Kohno, M.; Mukoyama, T.; Nishimuki, E.; et al. Aberrant expression of selenoproteins in the progression of colorectal cancer. *Cancer Lett.* **2008**, *259*, 218–230. [CrossRef] [PubMed]

16. Meplan, C.; Hughes, D.J.; Pardini, B.; Naccarati, A.; Soucek, P.; Vodickova, L.; Hlavata, I.; Vrana, D.; Vodicka, P.; Hesketh, J.E. Genetic variants in selenoprotein genes increase risk of colorectal cancer. *Carcinogenesis* **2010**, *31*, 1074–1079. [CrossRef] [PubMed]

17. Reszka, E.; Jablonska, E.; Gromadzinska, J.; Wasowicz, W. Relevance of selenoprotein transcripts for selenium status in humans. *Genes Nutr.* **2012**, *7*, 127–137. [CrossRef] [PubMed]

18. Flanagan, L.; Schmid, J.; Ebert, M.; Soucek, P.; Kunicka, T.; Liska, V.; Bruha, J.; Neary, P.; Dezeeuw, N.; Tommasino, M.; et al. Fusobacterium nucleatum associates with stages of colorectal neoplasia development, colorectal cancer and disease outcome. *Eur. J. Clin. Microbiol. Infect. Dis.* **2014**, *33*, 1381–1390. [CrossRef] [PubMed]

19. Winawer, S.J.; Zauber, A.G. The advanced adenoma as the primary target of screening. *Gastrointest. Endosc. Clin. N. Am.* **2002**, *12*, 1–9. [CrossRef]

20. Hlavata, I.; Mohelnikova-Duchonova, B.; Vaclavikova, R.; Liska, V.; Pitule, P.; Novak, P.; Bruha, J.; Vycital, O.; Holubec, L.; Treska, V.; et al. The role of ABC transporters in progression and clinical outcome of colorectal cancer. *Mutagenesis* **2012**, *27*, 187–196. [CrossRef] [PubMed]

21. Brynychova, V.; Hlavac, V.; Ehrlichova, M.; Vaclavikova, R.; Pecha, V.; Trnkova, M.; Wald, M.; Mrhalova, M.; Kubackova, K.; Pikus, T.; et al. Importance of transcript levels of caspase-2 isoforms s and l for breast carcinoma progression. *Future Oncol.* **2013**, *9*, 427–438. [CrossRef] [PubMed]

22. Soucek, P.; Anzenbacher, P.; Skoumalova, I.; Dvorak, M. Expression of cytochrome p450 genes in CD34+ hematopoietic stem and progenitor cells. *Stem Cells* **2005**, *23*, 1417–1422. [CrossRef] [PubMed]

23. Bustin, S.A. Why the need for qPCR publication guidelines?—The case for MIQE. *Methods* **2010**, *50*, 217–226. [CrossRef] [PubMed]

24. Hughes, D.J.; Duarte-Salles, T.; Hybsier, S.; Trichopoulou, A.; Stepien, M.; Aleksandrova, K.; Overvad, K.; Tjonneland, A.; Olsen, A.; Affret, A.; et al. Prediagnostic selenium status and hepatobiliary cancer risk in the European prospective investigation into cancer and nutrition cohort. *Am. J. Clin. Nutr.* **2016**, *104*, 406–414. [CrossRef] [PubMed]

25. Meplan, C.; Hesketh, J. The influence of selenium and selenoprotein gene variants on colorectal cancer risk. *Mutagenesis* **2012**, *27*, 177–186. [CrossRef] [PubMed]

26. Kipp, A.; Banning, A.; van Schothorst, E.M.; Meplan, C.; Schomburg, L.; Evelo, C.; Coort, S.; Gaj, S.; Keijer, J.; Hesketh, J.; et al. Four selenoproteins, protein biosynthesis, and Wnt signalling are particularly sensitive to limited selenium intake in mouse colon. *Molecular Nutr. Food Res.* **2009**, *53*, 1561–1572. [CrossRef] [PubMed]

27. Speckmann, B.; Steinbrenner, H. Selenium and selenoproteins in inflammatory bowel diseases and experimental colitis. *Inflamm. Bowel Dis.* **2014**, *20*, 1110–1119. [CrossRef] [PubMed]

28. Hesketh, J.; Meplan, C. Transcriptomics and functional genetic polymorphisms as biomarkers of micronutrient function: Focus on selenium as an exemplar. *Proc. Nutr. Soc.* **2011**, *3*, 1–9. [CrossRef] [PubMed]

29. Ren, B.; Liu, M.; Ni, J.; Tian, J. Role of selenoprotein F in protein folding and secretion: Potential involvement in human disease. *Nutrients* **2018**, *10*, 1619. [CrossRef] [PubMed]

30. Irons, R.; Tsuji, P.A.; Carlson, B.A.; Ouyang, P.; Yoo, M.H.; Xu, X.M.; Hatfield, D.L.; Gladyshev, V.N.; Davis, C.D. Deficiency in the 15-kDa selenoprotein inhibits tumorigenicity and metastasis of colon cancer cells. *Cancer Prev. Res.* **2010**, *3*, 630–639. [CrossRef] [PubMed]

31. Tsuji, P.A.; Carlson, B.A.; Naranjo-Suarez, S.; Yoo, M.H.; Xu, X.M.; Fomenko, D.E.; Gladyshev, V.N.; Hatfield, D.L.; Davis, C.D. Knockout of the 15 kDa selenoprotein protects against chemically-induced aberrant crypt formation in mice. *PLoS ONE* **2012**, *7*, e50574. [CrossRef] [PubMed]

32. Mork, H.; Al-Taie, O.H.; Bahr, K.; Zierer, A.; Beck, C.; Scheurlen, M.; Jakob, F.; Kohrle, J. Inverse mRNA expression of the selenocysteine-containing proteins GI-GPX and SePP in colorectal adenomas compared with adjacent normal mucosa. *Nutr. Cancer* **2000**, *37*, 108–116. [CrossRef] [PubMed]

33. Novoselov, S.V.; Kryukov, G.V.; Xu, X.M.; Carlson, B.A.; Hatfield, D.L.; Gladyshev, V.N. Selenoprotein H is a nucleolar thioredoxin-like protein with a unique expression pattern. *J. Biol. Chem.* **2007**, *282*, 11960–11968. [CrossRef] [PubMed]

34. Wang, N.; Chen, Y.; Yang, X.; Jiang, Y. Selenium-binding protein 1 is associated with the degree of colorectal cancer differentiation and is regulated by histone modification. *Oncol. Rep.* **2014**, *31*, 2506–2514. [CrossRef] [PubMed]

35. Hughes, D.J.; Fedirko, V.; Jones, J.S.; Méplan, C.; Schomburg, L.; Hybsier, S.; Riboli, E.; Hesketh, J.; Jenab, M. Association of selenoprotein and selenium pathway genetic variations with colorectal cancer risk and interaction with selenium status. In *Global Advances in Selenium Research from Theory to Application*, 1st ed.; Banuelos, G.S., Lin, Z.Q., Moraes, M.F., Guilherme, L.R.G., dos Reis, A.R., Eds.; Taylor & Francis: London, UK, 2015; pp. 53–54, ISBN 9781138027312.

36. Click, B.; Pinsky, P.F.; Hickey, T.; Doroudi, M.; Schoen, R.E. Association of colonoscopy adenoma findings with long-term colorectal cancer incidence. *JAMA* **2018**, *319*, 2021–2031. [CrossRef] [PubMed]

37. Combs, G.F., Jr. Biomarkers of selenium status. *Nutrients* **2015**, *7*, 2209–2236. [CrossRef] [PubMed]
38. Johnson, C.C.; Fordyce, F.M.; Rayman, M.P. Symposium on 'Geographical and geological influences on nutrition': Factors controlling the distribution of selenium in the environment and their impact on health and nutrition. *Proc. Nutr. Soc.* **2010**, *69*, 119–132. [CrossRef] [PubMed]

Article

Selenized Plant Oil Is an Efficient Source of Selenium for Selenoprotein Biosynthesis in Human Cell Lines

Jordan Sonet [1], Maurine Mosca [1], Katarzyna Bierla [1], Karolina Modzelewska [1],
Anna Flis-Borsuk [1], Piotr Suchocki [2], Iza Ksiazek [3], Elzbieta Anuszewska [3],
Anne-Laure Bulteau [1,4], Joanna Szpunar [1], Ryszard Lobinski [1] and Laurent Chavatte [1,5,6,7,]*

[1] CNRS/UPPA, Institute of Analytical and Physical Chemistry for the Environment and Materials (IPREM), UMR5254, Hélioparc, F-64053 Pau, France
[2] Department of Bioanalysis and Analysis of Drugs, Medical University of Warsaw, Żwirki i Wigury 81, 02-091 Warszawa, Poland
[3] Department of Biochemistry and Biopharmaceuticals, National Medicines Institute, Chelmska 30/34, PL-00-725 Warszawa, Poland
[4] Institut de Génomique Fonctionnelle de Lyon, IGFL, CNRS/ENS UMR5242, 69007 Lyon, France
[5] Centre International de Recherche en Infectiologie, CIRI, 69007 Lyon, France
[6] Institut national de la santé et de la recherche médicale (INSERM) Unité U1111, 69007 Lyon, France
[7] Centre national de la recherche scientifique (CNRS), Ecole Normale Supérieure (ENS) de Lyon, Université Claude Bernard Lyon 1 (UCBL1), Unité Mixte de Recherche 5308 (UMR5308), 69007 Lyon, France
* Correspondence: laurent.chavatte@ens-lyon.fr; Tel.: +33-4-72-72-86-24

Received: 13 June 2019; Accepted: 2 July 2019; Published: 4 July 2019

Abstract: Selenium is an essential trace element which is incorporated in the form of a rare amino acid, the selenocysteine, into an important group of proteins, the selenoproteins. Among the twenty-five selenoprotein genes identified to date, several have important cellular functions in antioxidant defense, cell signaling and redox homeostasis. Many selenoproteins are regulated by the availability of selenium which mostly occurs in the form of water-soluble molecules, either organic (selenomethionine, selenocysteine, and selenoproteins) or inorganic (selenate or selenite). Recently, a mixture of selenitriglycerides, obtained by the reaction of selenite with sunflower oil at high temperature, referred to as Selol, was proposed as a novel non-toxic, highly bioavailable and active antioxidant and antineoplastic agent. Free selenite is not present in the final product since the two phases (water soluble and oil) are separated and the residual water-soluble selenite discarded. Here we compare the assimilation of selenium as Selol, selenite and selenate by various cancerous (LNCaP) or immortalized (HEK293 and PNT1A) cell lines. An approach combining analytical chemistry, molecular biology and biochemistry demonstrated that selenium from Selol was efficiently incorporated in selenoproteins in human cell lines, and thus produced the first ever evidence of the bioavailability of selenium from selenized lipids.

Keywords: selenium; selenoprotein; selenized lipids; Selol; Gpx1; Gpx4; Txnrd1; Txnrd2; ICP-MS

1. Introduction

Selenocysteine is a rare amino acid which is vital for the function of selenoproteins, a unique group of genetically encoded proteins in redox reactions. Twenty-five selenoprotein genes have been identified in the human genome [1]. Selenoprotein synthesis depends upon Se intake, Se transport and Se conversion to selenocysteine (Sec) and its co-translational insertion into selenoproteins [2,3]. Several chemical forms, including selenomethionine, selenocysteine, selenite, and selenate, account for almost all the selenium in diets. All these forms are absorbed without regulation, and all have high bioavailability [4,5]. Among them, selenomethionine accounts for 90% of the total selenium in plants and is randomly incorporated into proteins at methionine positions. Free selenomethionine is easily available through the digestion of dietary proteins. Selenocysteine can also be a selenium source from animal proteins. However, the highly reactive free selenocysteine is maintained at very low concentrations in tissues. Although inorganic selenium compounds (selenite, and selenate) are less abundant than selenomethionine, they are an efficient source of selenium for selenoprotein biosynthesis. Selenate is reduced to selenite prior to its utilization as a selenium source. Then, selenite is reduced to selenide (HSe⁻), which is the central precursor for selenoprotein biosynthesis (see Figure 1). All selenocompounds need to be metabolized in selenide in order to be incorporated in genetically encoded selenoproteins. Additionally, among the selenometabolites, which are mostly excreted in urine, only a selenosugar (1β-methylseleno-N-acetyl-D-galactosamine) is, to a certain extent, bioavailable for selenoprotein biosynthesis when injected intravenously in animals [6].

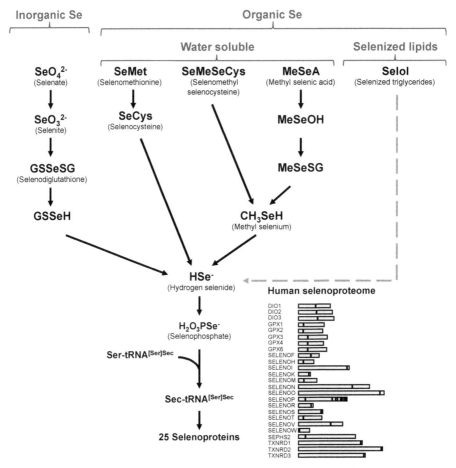

Figure 1. Representation of the various selenium metabolic pathways leading to selenoprotein synthesis. So far, every metabolic pathway towards selenoproteins converges to selenide. The dashed line suggest a putative metabolic pathway for selenium assimilation from selenized lipids.

Like for other essential trace elements, the selenium beneficial effect in cells follows a bell-shaped curve, a deficiency or an excess leading to toxicity and eventually lethality [7]. In cell lines, the window of benefit varies widely from one form of selenium to another and from one cell to another [5,8,9].

For example, in prostate cells exposed to selenite, the stimulation of selenoprotein expression is observed in the nanomolar range and toxicity starts at micromolar concentrations [10]. In contrast to that, selenate and selenomethionine are less toxic than selenite and the bell-shaped curve is shifted to higher concentrations. In liver cell lines, a comparative study with seven selenocompounds focusing on relative bioactivity and toxicity revealed highly variable behavior with regards to the chemical selenium form added to the culture media [4]. Again, at low doses, selenite showed a higher potential to activate selenoprotein synthesis than selenate; a similar shift in toxicity concentration was observed.

In genetically encoded selenoproteins, selenocysteine insertion occurs at UGA codons, which are normally used as a termination signal, and therefore follows a non-conventional mechanism [2,3,11–14]. To circumvent the reading of UGA as a stop signal, selenoprotein mRNAs have a Sec insertion sequence (SECIS) located in the 3′ untranslated region (UTR). This SECIS serves as a platform to recruit the recoding factors for selenocysteine insertion, to deliver the Sec-tRNA[Ser]Sec to the ribosomal A site. Interestingly, the SECIS element is necessary and sufficient to recode every in-frame UGA in a heterologous system. This feature has been particularly useful to characterize the functional determinants of the SECIS element in reporter constructs. Indeed a luciferase-based reporter construct that is stably expressed in HEK293 cells, in which the SECIS element of glutathione peroxidase-4 (Gpx4) is cloned downstream of the firefly luciferase containing an in frame UGA, was developed. When the UGA is read as a stop codon, the truncated luciferase is inactive, but when the UGA is read as a selenocysteine, the luciferase is fully active. The recoding efficiency is directly correlated to the luciferase activity measured in cellular extracts [15,16]. This reporter construct has been powerful to evaluate the variation of selenocysteine insertion efficiency in response to various stimuli which include selenium level variation, oxidative stress, and replicative senescence.

The search for bioavailable selenocompounds with lower toxicity than selenite drove research to design a selenized triglyceride, which is produced by the reaction of selenite with sunflower oil. This compound referred to as Selol is currently investigated for its potential anticancer properties and antineoplastic effects [17–21]. The reaction of sunflower oil with selenite generates a complex mixture of selenitriglycerides, which has been characterized recently by high performance liquid chromatography-inductively coupled plasma mass spectrometry (HPLC-ICP MS) and high performance liquid chromatography-electrospray ionization tandem mass spectrometry (HPLC ESI MS/MS). A total of 11 selenium-containing triglycerol derivatives have been identified where selenium is in the [Se(IV)] oxidation state [19], see Figure 2.

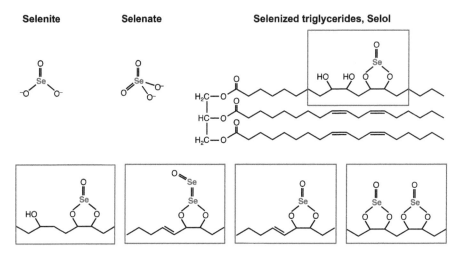

Figure 2. Chemical structures of selenite, selenate and Selol. In Selol, a complex mixture of at least 11 selenitriglyceride compounds have been characterized by high performance liquid chromatography-electrospray ionization tandem mass spectrometry (HLPC-ESI MS^n) [19].

However, the bioavailability of Selol for selenoprotein biosynthesis has not been investigated yet. The goal of this work was to investigate the assimilation of selenium from Selol and to compare it with that of inorganic species (selenite and selenate) in various cell lines, which are relevant for selenium physiological function. We are reporting here for the first time that a selenized lipid could be an efficient source of selenium for selenoprotein biosynthesis in HEK293 and LNCaP cell lines, which originate from embryonic kidney tissue and prostate adenocarcinoma, respectively.

2. Materials and Methods

2.1. Materials

The HEK293, LNCaP and PNT1A cells line used in this study were purchased from Life Technologies (Carlsbad, CA, USA, Cat.# R75007 and R71407), ATCC (Manassas, VA, USA, Cat.# CRL-1740) and SIGMA (Saint-Louis, MO, USA, Cat.# 95012614), respectively. The HEK293 cell line expressing Luc UGA/Gpx4 in a stable way was generated and validated in References [15,16,22]. Fetal calf serum (FCS), cell culture media and supplements were purchased from Life Technologies. Transferrin, insulin, 3,5,3'-triiodothyronine, hydrocortisone, sodium selenite and sodium selenate were purchased from Sigma-Aldrich (Saint-Louis, MO, USA). Antibodies were purchased from Abcam (Cambridge, UK) (glutathione peroxidase-1 (Gpx1), #ab108429; glutathione peroxidase-4 Gpx4, #ab125066; thioredoxin reductase-1 (TxnRD1), #ab124954) and Sigma-Aldrich (thioredoxin reductase-2 (Txnrd2), #HPA003323; Actin, #A1978). NuPAGE 4-12% bis–Tris polyacrylamide gels and MOPS SDS running buffer were purchased from Life Technologies. Antiprotease inhibitor cocktail was purchased from Thermo Fisher. Selol was synthesized at the Department of Bioanalysis and Drug Analysis at the Medical University of Warsaw, as described in Reference [23]. A micellar solution of Selol was used (based on lecithin, water and Selol) with a declared selenium concentration of 5% (w/v).

2.2. Cell Culture and Incubation with Different Forms of Selenium

HEK293 and LNCaP were grown and maintained in D-MEM, while RPMI medium was used for PNT1A. Media were supplemented with 10% FCS. Cells were cultivated at 37 °C and in a humidified atmosphere containing 5% CO_2. Since the selenium is endogenously provided by the FCS, the same lot number was kept throughout the experiments. Selenium concentration (194 nM) was determined by ICP MS in an FCS lot number (41G7530K) as reported in Reference [24]. Two different culture media, referred to as control (Ctrl) and depleted (Dpl) were used according to Reference [16]. Selenium concentration in the Ctrl medium was 19.4 nM. In the Dpl medium, we expected 3.9 nM selenium because 2% FCS was used instead of 10%. To cope with the decrease of growth factors in Dpl medium, 5 mg/L transferrin, 10 mg/L insulin, 100 pM 3,5,3'-triiodothyronine, and 50 nM hydrocortisone were added, as described and validated previously for selenoprotein expression studies.

The dose–response experiment with selenite, selenate and SELOL was performed with HEK293 cells stably expressing the Luc UGA/Gpx4 construct. Different media containing the respective concentrations of selenocompounds were prepared in Dpl media and used to cultivate the HEK293 cells for three days in a 10 cm (diameter) culture dish. After the treatment, cellular extracts were harvested with a volume of 300 µL of passive lysis buffer containing 25 mM Tris phosphate pH 7.8, 2 mM DTT, 2 mM EDTA, 1% Triton X100 and 10% glycerol. Then, protein concentrations were measured using the DC kit protein assay kit (Biorad, Hercules, CA, USA) in microplate assays using the microplate reader FLUOstar OPTIMA (BMG Labtech, Champigny-sur-Marne, France). Morphological evaluation was performed by phase-contrast microscopy using an Evos microscope (Ozyme, Saint-Cyr-l'École, France) with a 10X objective.

The comparative treatment of HEK293, LNCaP and PNT1A was performed in Ctrl medium supplemented or not with 100 nM or 100 µM of selenocompounds (selenite, selenate or Selol). Cells were grown for three days in 10 cm diameter culture dish.

2.3. Measure of Se Levels by ICP-MS

Flow injection–inductively coupled plasma mass spectrometry (FI-ICP-MS) methodology was used to measure total levels of selenium as reported in Reference [24]. Cellular extracts (200 µL) were mixed with 3 µL of nitric acid (70%) and left at room temperature overnight. Plasma conditions and detection parameters were optimized with 1 ppb Y, Li, Tl, and Ce in 2% of nitric acid. Signals of ^{77}Se, ^{78}Se, and ^{80}Se were acquired. A calibration curve was prepared from 5 ppm Se solution (calibration points: 1, 2.5, 5, 10, 25, 50, 100 ppb Se in 1 mL HNO_3, 1%). An Agilent (Santa Clara, CA, USA) 7500 ICP MS instrument fitted with a reaction/collision cell (H_2) was used. Every sample was analyzed by three consecutive injections of 50 µL.

2.4. Evaluation of Selenocysteine Insertion Efficiency

To analyze Sec insertion efficiency in HEK293, we used luciferase-based reporter constructs which were validated for UGA/Sec recoding in transfected cells [22,25]. Briefly, the minimal SECIS element from Gpx4 was cloned downstream of a luciferase coding sequence, which was modified to contain an in frame UGA codon at position 258 (Luc UGA/SECIS), as shown in Figure 3. HEK293 cells stably expressing Luc UGA/Gpx4 SECIS were previously generated and validated [15,16]. After growth in the presence of various concentrations of sodium selenite, cells were harvested and the cellular extracts were assayed for luciferase activity by chemiluminescence (luciferase assay systems, Promega, Madison, WI, USA), in triplicate using a microplate reader FLUOstar OPTIMA (BMG Labtech). We expressed the Sec insertion efficiency relative to the luciferase activity measured in Dpl conditions arbitrarily, setting it at 1.

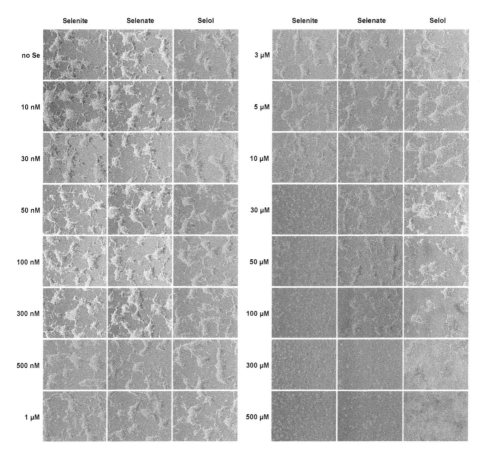

Figure 3. HEK293 morphology in response to various concentrations of three selenocompounds, namely selenite, selenate and Selol. Cells were grown for three days as described in materials and methods. Toxic effect of selenium is visible by cell detachment from the culture dish.

2.5. Protein Gels and Western Immunoblotting

Equal protein amounts (20 µg) were separated in Bis-Tris NuPAGE Novex Midi Gels and transferred onto nitrocellulose membranes using iBlot®dry blotting system (Life Technologies). Membranes were probed with primary antibodies (as indicated) and HRP-conjugated anti-rabbit or anti-mouse secondary antibodies (Sigma, #A9044 and #A6154, respectively). The chemiluminescence signal was detected using the ECL select Western blotting detection kit (GE Healthcare, Chicago, IL, USA) and the PXi 4 CCD camera (Ozyme). Image acquisition and data quantifications were performed with the Syngene softwares, GeneSys and Genetools, respectively.

3. Results

3.1. Comparison of the Selenium Uptake by HEK293 from Selol, Selenite and Selenate

The chemical structures of selenite, selenate and Selol are represented in Figure 2. Selenite and selenite differ in terms of their capacity to stimulate selenoprotein expression but also in terms of their active and toxic concentration ranges. In Selol, selenium thought to be at the oxidation state IV as in selenite, seems to be much less toxic than selenite in fibroblasts and prostate cells [23]. We investigated first the potential of these three selenocompounds to be taken up by HEK293 cells in a wide range of concentrations, by measuring the total selenium by ICP-MS. Dpl culture media containing 15 different concentrations of selenium (from 10 nM to 500 µM) in the form of selenite, selenate and Selol were prepared and used to grow HEK293 for three days. First, HEK293 cells were observed with a phase-contrast optical microscope and compared to the Dpl conditions without selenium (Figure 3). The toxicity of selenium was easily noticed using a microscope at the morphological level by the loss of cell attachment to the culture dish. As shown in Figure 3, this lack of attachment was observed at 30 µM and higher concentrations of selenite, but only at 300 µM and higher concentrations of selenate in the culture media.

Then, the cells (adherent and detached) grown in various selenium conditions were harvested and the cellular extracts were analyzed for the total selenium concentration by FI-ICP-MS. Selenium levels were normalized for the protein concentration, and plotted as a function of the selenium concentration in the culture medium, as illustrated in Figure 4. With the selenite concentrations in the culture medium of 100 µM and higher, the protein recovery was too low to allow the precise selenium detection (asterisk in Figure 4). Interestingly, we noticed that levels of selenium were increased at a maximum by three orders of magnitude to reach approximately 1 µg Se per mg of proteins. However, this increase was not a linear function of the selenium concentration in the growth medium. As shown in Figure 4, with selenate, a similar trend was observed, although shifted to higher concentrations. The loss of cell attachment started at a concentration of 300 µM in the culture media. Concerning Selol, a different behavior was observed with a linear relationship (at a log scale) between the assimilated selenium and the selenium added to the culture medium. With the maximal concentration of Selol tested here (500 µM), no sign of cell detachment was observed (Figure 3) and the Se concentration reached 12 µg per mg of proteins (Figure 4).

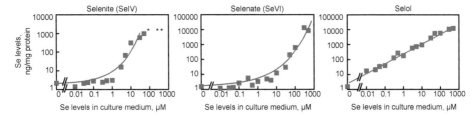

Figure 4. Measurement of selenium levels of HEK293 cells grown with increasing concentration of selenite, selenate or Selol in the culture media (from 10 nM up to 500 µM). Selenium levels are expressed relative to protein concentration. In several extracts, due to selenite toxicity, the protein recovery was too low to allow precise selenium detection and normalization, as indicated by an asterisk.

3.2. Selol is Able to Stimulate UGA Recoding as Selenocysteine in HEK293 Cells

The HEK293 cell line used in this study was genetically modified by a stably expressing luciferase-based reporter construct. The latter was validated previously to evaluate selenoprotein expression by measuring the efficiency of UGA recoding as selenocysteine [15,25,26]. Briefly, the minimal SECIS elements from Gpx4 3'UTR was cloned downstream of a luciferase coding sequence, which had been modified to contain an in frame UGA codon at position 258 (Luc UGA/SECIS), as shown in Figure 5. The luciferase activity was particularly sensitive to the Se supplementation of the culture, as shown previously with selenite [16]. The cellular extracts from cells grown in various concentrations of selenite, selenate and Selol were also evaluated for their luciferase activities. The enzymatic activities were normalized for the protein concentration and expressed relatively to the Dpl conditions, set as 1. As it has been mentioned before, an important stimulation of selenocysteine insertion was observed with low concentrations of selenite to reach a plateau at 100 nM up to 1 µM (see Figure 4). With higher doses of selenite in the culture medium, a rapid decrease of UGA recoding efficiency occurred to be almost undetectable at 10 µM. Interestingly, this decrease in selenocysteine insertion between 1 and 10 µM of selenite correlated with the massive increase in selenium levels (cf. graphs left panels in Figures 4 and 5). These data suggest that the level of selenium is tightly linked with the optimal selenocysteine insertion, and that overwhelming selenium levels seem to shut down the selenoprotein expression at a high dose. Alternatively, when selenate was used to supplement cell culture medium instead of selenite, a similar stimulation of UGA recoding efficiency was observed but shifted to higher concentrations. The plateau was reached at 1 μM of selenate and started decreasing at 100 μM. Again, when selenium levels and selenocysteine insertion efficiency were compared at high selenate concentrations (cf. Figures 4 and 5), we observed a similar inverse correlation as with selenite. These data confirm further that an excess of intracellular inorganic selenium could be detrimental to the cell integrity. Then, we investigated whether Selol was able to modulate the selenocysteine insertion efficiency. As shown in Figure 5, strong stimulation of UGA recoding activities occurred when Selol was added to the culture medium even at the lowest concentration tested (10 nM). A plateau of luciferase activity was reached at 100 nM and lasted up to 1 μM, similarly to selenite. Then, at higher concentrations, the recoding efficiency decreased slowly to reach the initial activity. These results demonstrate clearly for the first time the bioavailability of selenium bound to triglyceride for the selenocysteine insertion into proteins.

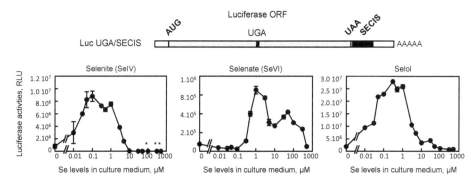

Figure 5. Evaluation of selenocysteine insertion efficiency in response to selenium supplementation in HEK293 cells. The luciferase activities from the stably expressing cell lines are measured from cells grown in increasing concentration of selenite, selenate or Selol in the culture media (from 10 nM up to 500 μM). Luciferase activities are expressed relative to protein concentration (relative luciferase unit, RLU). In several extracts, due to selenium toxicity, the protein recovery is too low to allow precise selenium detection and normalization, as indicated by an asterisk. A schematic of the luciferase construct is shown on top of histograms. AUG, start codon; UGA, selenocysteine codon; UAA, stop codon; SECIS, selenocysteine insertion sequence.

3.3. Selol Upregulates Selenoprotein Expression Also in LNCaP but not in PNT1A

Then, we investigated whether Selol was able to stimulate selenoprotein expression in prostate cell lines since this source of selenium was considered as a potential antineoplastic drug in prostate cancer. Here, we used LNCaP and PNT1A prostate cell lines, which are cancerous or immortalized, respectively. The LNCaP cells were previously found to be more sensitive than PNT1A to selenium-induced toxicity, both with selenite and Selol [23]. Here, we worked at similar selenium concentrations (i.e., 100 nM and 100 μM) with these cell lines as with HEK293 cells. As illustrated in Figures 6 and 7, the response of LNCaP to selenium supplementation was very similar to that of HEK293 cell lines; a strong stimulation of Gpx1 and Gpx4 was observed with 100 nM of selenite, and with 100 μM of selenate or Selol. Our data further support the bioavailability of Selol for selenoprotein biosynthesis. For the housekeeping selenoproteins, Selol (used at 100 nM or 100 μM) was also able to upregulate significantly Txnrd1 and Txnrd2 in LNCaP, similarly to selenite and selenate. On the other hand, PNT1A appeared poorly sensitive to selenium level variations. Interestingly, only the addition of 100 nM selenite to the culture medium had a moderate effect on Gpx1 and Gpx4, selenate and Selol being ineffective.

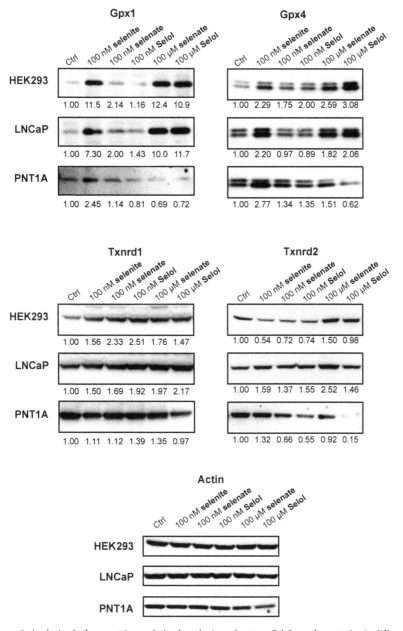

Figure 6. Analysis of selenoprotein regulation by selenite, selenate or Selol supplementation in different cell lines. HEK293, LNCaP and PNT1A cells were grown with Ctrl medium supplemented with either 100 nM or 100 μM of different selenium sources. Cellular extracts were analyzed for selenoprotein expression by western blots using specific antibodies. Protein levels were normalized over Actin levels and expressed relative to the Ctrl condition (no added selenium), set as 1.

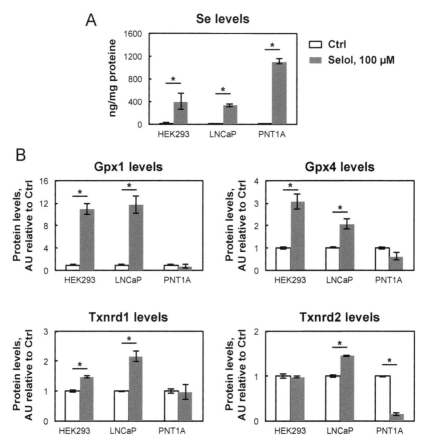

Figure 7. Comparison of Selol activities in three different human cell lines. (**A**) Selenium content in HEK293, LNCaP, PNT1A treated or not with 100 μM Selol. (**B**) Quantification of selenoprotein expression levels detected in Figure 6, normalized over the actin signal, in response to Selol supplementation. The significantly different (student test, $p < 0.01$) variations are indicated with an asterisk above the respective bars.

To confirm further that Selol has a cell line specific effect on selenoprotein expression, we verified that the differences observed between these cell lines did not result from a difference in selenium intake. For that, selenium levels were measured in the presence of 100 μM Selol (Figure 7, top histogram). We found that the different cell lines tested in our study were equally competent to take up selenium from Selol. In addition, our data suggest that the lack of selenoprotein upregulation by Selol in PNT1A may result from a difference in the capacity to efficiently metabolize Selol in an active precursor of Sec-tRNA[Ser]Sec, than in other cell lines such as HEK293 or LNCaP.

4. Discussion

In the environment, Se is present in many chemical forms and at variable concentrations. As selenium shares many chemical properties with sulfur, virtually any sulfur-containing molecule can have its analog with selenium instead [5,27]. This has been particularly observed in organisms such as higher plants and yeast that do not require selenium but are grown in a selenium-rich environment. The common strategy to detoxify this element is to incorporate selenium into organic molecules. However, to our knowledge, no selenized-lipid molecule has been observed so far in the wide diversity of the natural organic selenium species [27]. Among the selenodrugs that have been designed and synthesized, mostly for antioxidant and anticancer properties [28], Selol is a mixture of selenized triglycerides obtained by the reaction of Se_{+IV} with sunflower oil, in which at least 11 different selenium-containing triglycerol derivatives were characterized by mass spectrometry [19]. Here, we demonstrated that the selenium present in these compounds was efficiently bioavailable for human cell lines. It is widely accepted that the genetically encoded selenoproteins represent the biologically active form of selenium in mammalian cells. Among the twenty-five selenoproteins, many are involved in antioxidant defense, redox homeostasis and redox signaling. The selenoproteome is composed of five glutathione peroxidases (Gpx1-Gpx4, Gpx6), three thioredoxin reductases (Txnrd1-Txnrd3), one methionine sulfoxide reductase (MsrB1) and seven selenoproteins located in the endoplasmic reticulum (Dio2, SelenoF, SelenoK, SelenoM, SelenoN, SelenoS and SelenoT). However, about half of the selenoproteome remains without a precise function [3]. In the selenoproteome, Gpx1, Gpx4, Txnrd1 and Txnrd2 are ubiquitously expressed and react differently to the variation of selenium level. That is why we compared the potential of Selol to stimulate these selenoproteins with selenite and selenate, which are currently used for supplementation studies. Our main finding was that human cells use selenium from a mixture of selenized lipids for the synthesis of selenoproteins. One significant difference of Selol with other selenium compounds was the cell line specific activities. Indeed, Selol appeared to indifferently enter the three cell lines studied here but its use for selenoprotein synthesis was only efficient in HEK293 and LNCaP, but not in PNT1A which are non-cancerous prostate cell lines. This selectivity can come from a difference in lipid metabolism between the different cell lines studied in the present work. Indeed, as illustrated in Figure 2 and discussed in detail in Reference [5], every chemical selenium species that enters the cell has to be transformed into selenide to be further incorporated into selenoproteins. Three metabolic routes have been identified so far to generate hydrogen selenide, namely from selenite, selenocysteine, and methylselenol. How selenite is released from the lipids and whether all selenized triglycerides are equally competent for this release awaits further investigation. One possibility is that selenium is made bioavailable by lipolysis of the triglycerides followed by beta-oxidation of fatty acids.

Investigations about Selol mostly concern its anticancer activities [17–21,23,29], very little work has been done on antioxidant activities and potential use for not cancerous cells. The present study demonstrated the high level of selenium assimilation from this mixture of selenized-lipids and confirmed the lower toxicity in comparison to other selenocompounds. The cellular selectivity for insertion into selenoproteins has to be extended to other cellular models, either cancerous or not, to get a better picture of whether this new compound can be used as a dietary supplement.

5. Conclusions

This study is, to our knowledge, the first report for assimilation of Se from lipids into selenoproteins. Whether selenized lipids are present in the human food diet is an open issue due to the inherent experimental detection and characterization difficulties. Clearly, for the cell lines tested so far, Selol is much less toxic than selenite and selenate. Now being tested for antineoplastic activity, it may be considered as a potential nutritional source of selenium for deficient populations. Similarly to selenite and selenate, Selol is able to upregulate selenoprotein synthesis by following a precise prioritization of selenium use, also referred to as selenoprotein hierarchy [2,3]. Selol has a cell line specific capacity to upregulate selenocysteine insertion. This selectivity can be due to a cell line specific ability to metabolize Selol to make selenium bioavailable, most probably by lipolysis and beta-oxidation of fatty acid pathways. Despite similar uptake, the metabolism of Selol in selenide is different in PNT1A from HEK293 or LNCaP cells.

Author Contributions: Conceptualization, J.S. (Jordan Sonet), A.F.-B., A.-L.B., J.S. (Joanna Szpunar), R.L. and L.C.; Formal analysis, J.S. (Jordan Sonet), M.M., and L.C.; Funding acquisition, R.L. and L.C.; Investigation, J.S. (Jordan Sonet), M.M., K.B., K.M., A.-L.B., R.L. and L.C.; Methodology, J.S. (Jordan Sonet), M.M., K.B., K.M., A.F.-B., P.S., I.K., E.A., A.-L.B., and L.C.; Project administration, J.S. (Jordan Sonet), R.L. and L.C.; Resources, J.S. (Jordan Sonet),

A.F.-B., P.S., I.K., E.A., and L.C.; Supervision, J.S. (Joanna Szpunar), R.L. and L.C.; Validation, J.S. (Jordan Sonet); Visualization, J.S. (Jordan Sonet), M.M. and L.C.; Writing—original draft, A.-L.B., R.L. and L.C.; Writing—review and editing, J.S. (Jordan Sonet), M.M., K.B., K.M., A.F.-B., P.S., I.K., E.A., A.-L.B., J.S. (Joanna Szpunar), R.L. and L.C.

Funding: This research was funded by Agence Nationale de la Recherche (ANR-09-BLAN-0048, R.L. and L.C.), Institut National de la Santé et de la Recherche Médicale: INSERM (L.C.), Centre National de la Recherche Scientifique: CNRS (ATIP program, L.C.), Ecole normale supérieure de Lyon: ENS lyon (Emerging Project, L.C.). J.S. (Jordan Sonet) was a recipient of a Ph.D. fellowship from the Université de Pau et des Pays de l'Adour (UPPA).

Acknowledgments: The contributions of the Good Life Foundation (Syracuse, NY) and ERASMUS program are acknowledged.

Conflicts of Interest: The authors declare no conflict of interest.

References

1. Kryukov, G.V.; Castellano, S.; Novoselov, S.V.; Lobanov, A.V.; Zehtab, O.; Guigo, R.; Gladyshev, V.N. Characterization of mammalian selenoproteomes. *Science* **2003**, *300*, 1439–1443. [CrossRef] [PubMed]
2. Bulteau, A.-L.; Chavatte, L. Update on selenoprotein biosynthesis. *Antioxid. Redox Signal.* **2015**, *23*, 775–794. [CrossRef] [PubMed]
3. Vindry, C.; Ohlmann, T.; Chavatte, L. Translation regulation of mammalian selenoproteins. *Biochim. Biophys. Acta Gen. Subj.* **2018**, *1862*, 2480–2492. [CrossRef] [PubMed]
4. Hoefig, C.S.; Renko, K.; Kohrle, J.; Birringer, M.; Schomburg, L. Comparison of different selenocompounds with respect to nutritional value vs. toxicity using liver cells in culture. *J. Nutr. Biochem.* **2011**. [CrossRef] [PubMed]
5. Vindry, C.; Ohlmann, T.; Chavatte, L. Selenium metabolism, regulation, and sex differences in mammals. In *Selenium, Molecular and Integrative Toxicology*; Michalke, B., Ed.; Springer Nature: Berlin/Heidelberg, Germany, 2018; pp. 89–107.
6. Suzuki, K.T.; Somekawa, L.; Suzuki, N. Distribution and reuse of 76Se-selenosugar in selenium-deficient rats. *Toxicol. Appl. Pharmacol.* **2006**, *216*, 303–308. [CrossRef] [PubMed]
7. Sonet, J.; Bulteau, A.-L.; Chavatte, L. Selenium and Selenoproteins in Human Health and Diseases. In *Metallomics: Analytical Techniques and Speciation Methods*; Michalke, B., Ed.; Wiley-VCH Verlag GmbH & Co. KGaA: Weinheim, Germany, 2016; pp. 364–381, Chapter 13. [CrossRef]
8. Touat-Hamici, Z.; Legrain, Y.; Sonet, J.; Bulteau, A.-L.; Chavatte, L. Alteration of selenoprotein expression during stress and in aging. In *Selenium: Its Molecular Biology and Role in Human Health*, 4th ed.; Hatfield, D.L., Tsuji, P.A., Gladyshev, V.N., Eds.; Springer Science+Business Media, LLC: New York, NY, USA, 2016; pp. 539–551.
9. Touat-Hamici, Z.; Bulteau, A.L.; Bianga, J.; Jean-Jacques, H.; Szpunar, J.; Lobinski, R.; Chavatte, L. Selenium-regulated hierarchy of human selenoproteome in cancerous and immortalized cells lines. *Biochim. Biophys. Acta Gen. Subj.* **2018**. [CrossRef] [PubMed]
10. Rebsch, C.M.; Penna, F.J., 3rd.; Copeland, P.R. Selenoprotein expression is regulated at multiple levels in prostate cells. *Cell Res.* **2006**, *16*, 940–948. [CrossRef] [PubMed]
11. Allmang, C.; Wurth, L.; Krol, A. The selenium to selenoprotein pathway in eukaryotes: More molecular partners than anticipated. *Biochim. Biophys. Acta* **2009**, *1790*, 1415–1423. [CrossRef] [PubMed]
12. Driscoll, D.M.; Copeland, P.R. Mechanism and regulation of selenoprotein synthesis. *Annu. Rev. Nutr.* **2003**, *23*, 17–40. [CrossRef]
13. Hatfield, D.L.; Gladyshev, V.N. How selenium has altered our understanding of the genetic code. *Mol. Cell Biol.* **2002**, *22*, 3565–3576. [CrossRef]
14. Berry, M.J.; Tujebajeva, R.M.; Copeland, P.R.; Xu, X.M.; Carlson, B.A.; Martin, G.W., 3rd.; Low, S.C.; Mansell, J.B.; Grundner-Culemann, E.; Harney, J.W.; et al. Selenocysteine incorporation directed from the 3'UTR: Characterization of eukaryotic EFsec and mechanistic implications. *Biofactors* **2001**, *14*, 17–24. [CrossRef] [PubMed]
15. Touat-Hamici, Z.; Legrain, Y.; Bulteau, A.-L.; Chavatte, L. Selective up-regulation of human selenoproteins in response to oxidative stress. *J. Biol. Chem.* **2014**, *289*, 14750–14761. [CrossRef] [PubMed]
16. Latreche, L.; Duhieu, S.; Touat-Hamici, Z.; Jean-Jean, O.; Chavatte, L. The differential expression of glutathione peroxidase 1 and 4 depends on the nature of the SECIS element. *RNA Biol.* **2012**, *9*, 681–690. [CrossRef] [PubMed]

17. Dominiak, A.; Wilkaniec, A.; Jesko, H.; Czapski, G.A.; Lenkiewicz, A.M.; Kurek, E.; Wroczynski, P.; Adamczyk, A. Selol, an organic selenium donor, prevents lipopolysaccharide-induced oxidative stress and inflammatory reaction in the rat brain. *Neurochem. Int.* **2017**, *108*, 66–77. [CrossRef] [PubMed]

18. Sliwka, L.; Wiktorska, K.; Suchocki, P.; Milczarek, M.; Mielczarek, S.; Lubelska, K.; Cierpial, T.; Lyzwa, P.; Kielbasinski, P.; Jaromin, A.; et al. The Comparison of MTT and CVS Assays for the Assessment of Anticancer Agent Interactions. *PLoS ONE* **2016**, *11*, e0155772. [CrossRef] [PubMed]

19. Bierla, K.; Flis-Borsuk, A.; Suchocki, P.; Szpunar, J.; Lobinski, R. Speciation of Selenium in Selenium-Enriched Sunflower Oil by High-Performance Liquid Chromatography-Inductively Coupled Plasma Mass Spectrometry/Electrospray-Orbitrap Tandem Mass Spectrometry. *J. Agric. Food Chem.* **2016**, *64*, 4975–4981. [CrossRef] [PubMed]

20. Flis, A.; Suchocki, P.; Krolikowska, M.A.; Suchocka, Z.; Remiszewska, M.; Sliwka, L.; Ksiazek, I.; Sitarz, K.; Sochacka, M.; Hoser, G.; et al. Selenitetriglycerides-Redox-active agents. *Pharmacol. Rep.* **2015**, *67*, 1–8. [CrossRef] [PubMed]

21. Sochacka, M.; Giebultowicz, J.; Remiszewska, M.; Suchocki, P.; Wroczynski, P. Effects of Selol 5% supplementation on the activity or concentration of antioxidants and malondialdehyde level in the blood of healthy mice. *Pharmacol. Rep.* **2014**, *66*, 301–310. [CrossRef] [PubMed]

22. Chavatte, L.; Brown, B.A.; Driscoll, D.M. Ribosomal protein L30 is a component of the UGA-selenocysteine recoding machinery in eukaryotes. *Nat. Struct. Mol. Biol.* **2005**, *12*, 408–416. [CrossRef]

23. Ksiazek, I.; Sitarz, K.; Anuszewska, E.; Dudkiewicz-Wilczynska, J.; Rolson, M.; Koronkiewicz, M.; Suchocki, P. Toxicity studies of selol—An organic selenium (iv) compound- in vitro research. *Int. J. Pharm. Pharm. Sci.* **2014**, *6*, 264–269.

24. Vacchina, V.; Dumont, J. Total Selenium Quantification in Biological Samples by Inductively Coupled Plasma Mass Spectrometry (ICP-MS). *Methods Mol. Biol.* **2018**, *1661*, 145–152. [CrossRef] [PubMed]

25. Latreche, L.; Jean-Jean, O.; Driscoll, D.M.; Chavatte, L. Novel structural determinants in human SECIS elements modulate the translational recoding of UGA as selenocysteine. *Nucleic Acids Res.* **2009**, *37*, 5868–5880. [CrossRef] [PubMed]

26. Legrain, Y.; Touat-Hamici, Z.; Chavatte, L. Interplay between selenium levels, selenoprotein expression, and replicative senescence in WI-38 human fibroblasts. *J. Biol. Chem.* **2014**, *289*, 6299–6310. [CrossRef] [PubMed]

27. Bierla, K.; Szpunar, J.; Lobinski, R. Biological Selenium Species and Selenium Speciation in Biological Samples. In *Selenium: Its Molecular Biology and Role in Human Health*, 4th ed.; Hatfield, D.L., Tsuji, P.A., Gladyshev, V.N., Eds.; Springer Science+Business Media, LLC: New York, NY, USA, 2016; pp. 413–424. [CrossRef]

28. Ramoutar, R.R.; Brumaghim, J.L. Antioxidant and anticancer properties and mechanisms of inorganic selenium, oxo-sulfur, and oxo-selenium compounds. *Cell Biochem. Biophys.* **2010**, *58*, 1–23. [CrossRef] [PubMed]

29. Ksiazek, I.; Sitarz, K.; Roslon, M.; Anuszewska, E.; Suchocki, P.; Wilczynska, J.D. The influence of Selol on the expression of oxidative stress genes in normal and malignant prostate cells. *Cancer Genom. Proteom.* **2013**, *10*, 225–232.

Communication

Selenium and Health: An Update on the Situation in the Middle East and North Africa

Sohayla A. Z. Ibrahim [1], Abdelhamid Kerkadi [2] and Abdelali Agouni [1,*]

[1]   Department of Pharmaceutical Sciences, College of Pharmacy, QU health, Qatar University, P.O. Box 2713,
      Doha, Qatar
[2]   Department of Nutrition, College of Health Sciences, QU health, Qatar University, P.O. Box 2713, Doha, Qatar
*   Correspondence: aagouni@qu.edu.qa; Tel.: +974-4403-5610

Received: 29 May 2019; Accepted: 19 June 2019; Published: 27 June 2019

Abstract: Selenium (Se) is an important trace element that should be present in the diet of all age groups to provide an adequate intake. Se is incorporated in 25 known selenoproteins, which mediate the biological effects of Se including, immune response regulation, maintenance of thyroid function, antioxidant defense, and anti-inflammatory actions. A balanced intake of Se is critical to achieve health benefits because depending on its status, Se has been found to play physiological roles or contribute to the pathophysiology of various diseases including, neurodegenerative diseases, diabetes, cancer, and cardiovascular disorders. Se status and intake are very important to be known for a specific population as the levels of Se are highly variable among different populations and regions. In the Middle East and North African (MENA) region, very little is known about the status of Se. Studies available show that Se status is widely variable with some countries being deficient, some over sufficient, and some sufficient. This variability was apparent even within the same country between regions. In this review, we summarized the key roles of Se in health and disease and discussed the available data on Se status and intake among countries of the MENA region.

Keywords: selenium; chronic diseases; Middle East; North Africa

---

1. Introduction

Selenium (Se) is a semi-solid metal that was discovered in 1817 as a byproduct of sulphuric acid synthesis. It belongs to group 6 of the periodic table with an atomic number of 34. It is a red colored powder; however, in vitreous form it is observed as black, and in crystalline form it is observed as metallic gray. Se exists in many different oxidation states including 2+, 4+, 6+, and 2- [1]. Se is present in small amounts in food mostly in organic forms (selenomethionine and selenocysteine) and rarely in inorganic forms (selenate and selenite) [2]. Selenomethionine is usually derived from animal sources in addition to cereal products grown in areas where the soil is rich in Se whereas selenocysteine is only obtained from animal sources [3]. Inorganic forms are the major sources of Se incorporated into dietary supplements [2]. Se concentrations in the soil where plants are grown or where animals are raised are important indicators of Se intake in a country leading to tremendous variations among different countries in Se status [2,4]. The most common sources of Se are Brazil nuts, cereals, offal, fish, eggs, poultry, and vegetables [3–5].

Se is a necessary trace element that should be present in the diet of all age groups. The Recommended Dietary Allowance (RDA) of Se is generally based on the optimum amount that can maximize the activity of selenoprotein glutathione peroxidase [2–4]. This value was estimated to be 55 μg/day for both males and females [2–4,6]. The Estimated Average Requirement (EAR) was found to be around 45 μg/day for 19–50-year-old men and women and the tolerable upper intake level (UL) is around 400 μg/day [3,4]. However, the optimal amount required to achieve maximum health benefits is unknown [3]. The World Health Organization (WHO) recommends that the optimum serum concentration of Se for healthy adults is 39.5–194.5 ng/ml and that the concentration that maximizes glutathione peroxidase activity is between 70 and 90 ng/ml [7,8].

Very limited data exist on Se toxicity in humans; however, animal studies are available on chronic Se poisoning [4]. It was reported that an intake of around 800 μg/day of Se would cause no observed adverse effect level (NOAEL), 1,540 to 1,600 μg/day of Se would result in low observed adverse effect level (LOAEL) and 5,000 μg/day of Se is the toxic level where selenosis is expected to happen [3]. LOAEL is associated with symptoms of hair and nail brittleness and loss, garlic like breath odor, gastrointestinal disturbances, and fatigue [3,4]. Selenosis can be associated with serious respiratory, renal, and cardiac complications [3].

Se status can be assessed through blood, urine, nails, and hair levels. Blood specimens can be either whole blood, erythrocytes, serum, or plasma. Erythrocytes, hair, and nails reflect long-term status opposing to plasma or serum which reflect short-term levels. It is challenging to compare Se levels across different countries due to the wide variations in methodologies of Se detection among different laboratories [9,10].

2. Selenium: A Focus on Biology

2.1. Selenoproteins as Mediators of Se Actions

Se is an essential nutrient to human biology. The beneficial roles of Se to human health and its requirement for life have been known for several decades. Se exists in 25 known selenoproteins, some of which have important known biological functions [11]. The critical role of Se in human health is particularly underscored by being the sole trace element specified in the genetic code as selenocysteine, today recognized as the 21st amino acid, which has its proper genetic code, specific biosynthesis pathway, and insertion mechanism. This amino acid exists in the active site of many selenium-dependent enzymes that are responsible for multiple essential functions in the human body [1,12,13]. In this case, Se acts as a redox center. For example, thioredoxin reductase, a selenoenzyme, participates in DNA synthesis by reducing nucleotides which then helps in controlling the redox state intracellularly [3]. Another example is iodothyronine deiodinases which activates thyroid hormone [1]. Other selenoproteins have different functions including anti-inflammatory and antioxidant effects; however, some of those functions are yet to be identified. The biological effects of Se also include its participation in immune responses and thyroid function whereas its serum level can also be linked with risk of diabetes, cardiovascular disease (CVD), and cancer [5]. Table 1 summarizes the main functions and significance of the different known selenoproteins.

Table 1. Functions and significance of selenoproteins

Selenoprotein	Symbol	Function and Significance
Glutathione peroxidase 1	GPX1	Reduces cellular H_2O_2. Overexpression of GPX1 increases risk of diabetes.
Glutathione peroxidase 2	GPX2	Reduces peroxide in gut. GPX1/GPX2 double-knockout mice develop intestinal cancer; one allele of GPX2 added back confers protection.
Glutathione peroxidase 3	GPX3	Reduces peroxide in blood. Important for cardiovascular protection, perhaps through modulation of nitric oxide levels; antioxidant in thyroid gland.
Glutathione peroxidase 4	GPX4	Reduces phospholipid peroxide. Genetic deletion is embryonic lethal; GPX4 acts as crucial antioxidant, and sensor of oxidative stress and pro-apoptotic signals; required for spermatozoa function.
Glutathione peroxidase 6	GPX6	Importance unknown.

Table 1. *Cont.*

Selenoprotein	Symbol	Function and Significance
Iodothyronine deiodinase 1	DIO1	Important for systemic active thyroid hormone levels.
Iodothyronine deiodinase 2	DIO2	Important for local active thyroid hormone levels.
Iodothyronine deiodinase 3	DIO3	Inactivates thyroid hormone.
Thioredoxin reductase 1	TXNRD1	Reduction of cytosolic thioredoxin. Genetic deletion is embryonic lethal.
Thioredoxin reductase 2	TXNRD2	Reduction of mitochondrial thioredoxin. Genetic deletion is embryonic lethal.
Thioredoxin-glutathione reductase	TXNRD3	Reduction of thioredoxin, testes-specific expression.
Selenoprotein H	SELENOH	Involved in transcription. Essential for cell survival and antioxidant defense in Drosophila.
Selenoprotein I	SELENOI	Possibly involved in phospholipid biosynthesis.
Selenoprotein K	SELENOK	Involved in calcium flux in immune cells and endoplasmic reticulum (ER)-associated degradation.
Selenoprotein M	SELENOM	Thiol-disulfide oxidoreductase localized in the ER. Possibly involved in protein folding.
Selenoprotein F	SELENOF	Thiol-disulfide oxidoreductase localized in the ER. Possibly involved in protein folding.
Selenoprotein N	SELENON	Potential role in early muscle development. Mutations lead to multiminicore disease and other myopathies.
Selenoprotein O	SELENOO	Potential redox function, but importance remains unknown.
Selenoprotein P	SELENOP	Se transport to tissues particularly brain and testis. It also functions as intracellular antioxidant in phagocytes. Knockout leads to neurological problems and male sterility.
Methionine-R-sulfoxide reductase 1	MSRB1	Functions as a methionine sulfoxide reductase and MSRB1 knockouts show mild damage to oxidative insult.
Selenoprotein S	SELENOS	Transmembrane protein found in plasma membrane and ER. Reduces ER stress.
Selenoprotein T	SELENOT	ER protein involved in calcium mobilization.
Selenoprotein V	SELENOV	Testes-specific expression, potential role in male reproduction.
Selenoprotein W	SELENOW	Potential antioxidant role, perhaps important in muscle growth.
Selenophosphate synthetase 2	SEPHS2	Involved in the synthesis of all selenoproteins.

2.2. Synthesis and Co-Translational Incorporation of Selenocysteine in Humans

The cycle of selenocysteine synthesis and incorporation in humans begins with the attachment of L-serine to non-cognate tRNA^{Sec} by seryl-tRNA synthetase, an error that is not edited. This seryl group gets phosphorylated by O-phosphoseryl-tRNA^{Sec} kinase (PSTK) resulting in phosphoseryl (Sep)-tRNA^{Sec} intermediate. Finally, this Sep-tRNA^{Sec} is converted into selenocysteinyl (Sec)-tRNA^{Sec} through O-phosphoseryl-tRNA^{Sec}:selenocysteinyl-tRNA^{Sec} synthase (SepSecS) in the presence of selenophosphate and pyridoxal phosphate. Selenophosphate is the main Se donor in humans synthesized from adenosine triphosphate (ATP) and selenide, the products of selenocysteine degradation, by selenophosphate synthetase (SPS2). A specialized elongation factor (EFsec) then delivers Sec-tRNA^{Sec} to the ribosome. SElenoCysteine Insertion Sequence (SECIS), located in the 3'-UTR, helps localize EFsec and Sec-tRNA^{Sec} near the translation site and forms a stem loop structure which is necessary for decoding of selenocysteine UGA codon and its insertion into the nascent protein [14–17]. The cycle of selenocysteine synthesis and incorporation into selenoproteins in humans is summarized in Figure 1.

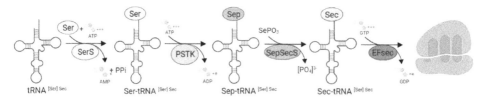

Figure 1. Synthesis and co-translational incorporation of selenocysteine into selenoproteins in humans.

3. Selenium in Health and Disease: The Importance of a Balanced Intake

Se depending on its status has been reported to play crucial roles in normal physiology or contribute the pathophysiology of various diseases. Se has antioxidant and anti-inflammatory actions and therefore various studies have assessed the impact of Se status on conditions characterized by inflammation and oxidative stress including, neurodegenerative diseases, diabetes, cancer, and cardiovascular disorders [18].

3.1. Pregnancy

During pregnancy, the demand for multiple nutrients increase owing to many physiological changes, resulting in deficiencies of many vitamins and minerals [9,19]. Those deficiencies might sometimes lead to complications to the fetus or to the mother. Se levels were found to be significantly reduced during pregnancy [9,19]. During the first trimester, GPx activity was found to be reduced significantly and this low level was maintained throughout the second and third trimesters with a slight increase during delivery [20]. Subsequently, in the second and third trimesters, Se levels were noted to be significantly reduced with further reduction during delivery [20]. Due to this increased demand, the RDA for Se during pregnancy was increased from 55 µg/day to 60 µg/day [19,21]. Se deficiencies in pregnant women might have adverse effects on the developing fetus, particularly the nervous system. A case-control study of pregnant women found a positive correlation between low Se levels and the occurrence of neural tube defects [22]. However, it is important to note that defects in the nervous system development are affected by multiple other factors and not solely by Se deficiencies. Se deficiencies during pregnancy were also found to cause oxidative stress partly resulting in miscarriages, pre-term deliveries, intrauterine growth retardations, preeclampsia, thyroid dysfunctions, gestational diabetes, and cholestasis [9,23]. Therefore, it is important to maintain optimal Se levels during pregnancy by increasing Se intake to meet the increased demand.

3.2. Diabetes Mellitus

The relationship between Se and diabetes was investigated in multiple studies with contradicting results [18]. Those studies were conducted initially following the assumption that Se is likely to exert protective effects against diabetes due to its antioxidant actions. In support of this assumption, Se was found to delay the onset and progression of diabetes. Se was also found to act as an insulin mimetic when in the form of selenite. Moreover, several in vitro and in vivo studies have suggested that Se plays an important role in the regulation of glucose homeostasis [18,24]. Human studies have also investigated the impact of Se on diabetes yet with conflicting conclusions. Both low and high levels of Se have been shown to be associated with a risk of diabetes [2]. The role of Se in diabetes was assessed in few randomized control trials (RCT). The first large RCT was the Nutritional Prevention of Cancer (NPC) trial which aimed at assessing the effect of 200 µg of daily Se supplementation on reducing the risk of skin cancer compared to placebo [25]. A secondary analysis of the data after a follow-up of 7 years suggested a statistically significant increase in the incidence of type 2 diabetes among those who received Se supplementation compared to placebo [26]. In contrast, the Se and Vitamin E Cancer Prevention Trial (SELECT) found no statistically significant increase in the incidence of type 2 diabetes after Se supplementation [27]. The third trial is the Se and Celecoxib (Sel/Cel) Trial.

The intervention was a randomized, placebo controlled trial of Se (200 µg/day) and cyclooxygenase 2 selective inhibitor, celecoxib (400 mg, once per day), alone or in combination, for the prevention of colorectal adenoma in men and women. After exclusion of participants who had type 2 diabetes prior to the trial, 1640 participants were assessed for the presence of type 2 diabetes during the follow-up period. Authors found that 31 participants randomized to Se and 25 randomly assigned placebo were diagnosed with type 2 diabetes; however, despite a higher number of cases in the Se arm, the difference was not statistically significant. Interestingly, the study showed a statistically significant increase in the risk of type 2 diabetes in elderly individuals, which suggests that with old age supplementation with Se may increase the risk of type 2 diabetes [28]. These findings mirror the controversy on whether higher Se concentrations could increase the risk of type 2 diabetes or not.

Observational studies have also shown a non-linear association between Se and type 2 diabetes. A recent, comprehensive meta-analysis that was conducted in 2019 among 20 observational studies concluded that high Se levels are significantly associated with type 2 diabetes (pooled odds ratios, 1.88; 95% confidence interval, 1.44 to 2.45). However, significant heterogeneity was observed within different studies and the analysis of the funnel plot revealed significant publication bias. Subgroup analysis was also conducted according to the method of Se measurement in each study. Significant association was shown among studies that used blood, diet, and urine as specimens for Se level detection but not in studies that used nails as a sample for Se measurement. Sensitivity analysis was conducted in addition to the trim and fill analysis to account for the heterogeneity and the publication bias, yet the results were consistent upon all adjustments [18]. Another non-linear dose–response analysis confirmed that both levels below and above the physiological range are potential risk factors for diabetes which is consistent with other review articles that suggested a U-shaped relationship between Se and diabetes. Positive associations were detected among patients with low Se levels (<100 µg/l) and patients with high Se levels (>130 µg/l) [29].

3.3. Cardiovascular Disease (CVD)

Se as an antioxidant was assumed to play an essential role in protecting against atherosclerotic events and CVD [30]. Se deficiency was directly linked to Keshan disease, an endemic cardiomyopathy that is found in regions where Se intake is very low [31]. Multiple studies were carried out to assess the impact of Se on CVD risk. Both low and high levels of Se were found to adversely affect heart function [2]. In Eastern USA, a post hoc analysis of the NPC trial was conducted with a follow-up of 7.6 years to study the effect of Se supplementation on the prevention of CVD. The incidence of myocardial infarction, total cerebrovascular accidents and CVD were assessed, and the results revealed that there is no overall benefit of supplementing 200 µg/day of Se to prevent CVD [26]. The SELECT trial did not also find any significant effect of Se on overall cardiovascular events [27]. A meta-analysis of RCTs showed that oral Se did not have an overall effect on cardiovascular events regardless of Se formulation or dose [32]. Observational studies have shown conflicting conclusions. A meta-analysis of observational studies concluded a nonlinear, U-shaped relationship, between Se levels and CVD risk. The suggested range of Se concentrations that results in significant beneficial effects against CVD is from 55 to 145 µg/l [32].

3.4. Cancer

The role of Se in cancer prevention was extensively studied in various cancer types. Se exerts its action through the antioxidant activity of selenoproteins in addition to several other mechanisms [33]. However, some laboratory studies suggested that Se can promote malignant cell transformation and progression [34]. Therefore, whether Se promotes or prevents cancer is not fully clear. Various RCTs investigated Se supplementation for cancer prevention. The first RCT is the NPC trial that aimed at assessing the influence of Se on the development of non-melanoma skin cancer. The NPC trial concluded that Se does not exert a chemoprotective effect against the recurrence of non-melanoma skin cancer, yet it significantly reduced the risk for all cancers, and for esophageal, prostate, colorectal, and lung cancers [25]. Another smaller trial among patients who received organ transplants found an unexpected increase in the incidence of non-melanoma skin cancer as a secondary outcome in Se-supplemented group [35]. The SELECT trial was a pivotal study that was carried out in male general population without high risk of prostate cancer to assess the incidence of prostate cancer among Se-supplemented participants compared to placebo. The study found no significant difference in prostate cancer incidence between Se-supplemented and placebo groups. It also detected no difference in overall cancer risk or any other cancer types [27]. Moreover, three recent RCTs evaluated Se impact on participants with high risk for prostate cancer and all showed no beneficial effect of Se on the incidence of cancer [36–38]. Additionally, a trial among women with high risk of breast cancer due to mutations to breast cancer type 1 (BRCA1) susceptibility gene showed increased risk of all cancers and primary breast cancer in Se-supplemented arm compared to placebo [39]. Observational studies have also assessed Se effect on cancer risk with strongly conflicting results. A recent meta-analysis of observational cohort studies showed an overall lower risk of cancer among participants with the highest baseline exposure levels, yet the included studies are subject to a substantial risk of bias [40].

4. Regional Variability of Se Status: A Focus on the Situation in the Middle Eastern and North Africa (MENA) Region

The status of Se across the MENA region is widely variable with some countries deficient, some over sufficient and some sufficient in addition to several countries with unknown Se status. This variability was apparent even between provinces from the same country. These tremendous variations may be linked to the varying concentration of Se in the soil were food consumed is gown. We summarized in this review the available evidence on Se status and intake among countries of the MENA region. It is important to note that most of the studies that assessed Se concentrations in biological samples were case-control studies with small sample sizes that cannot be generalized to the whole population. Few studies addressed Se status in a small number of countries with total paucity of information in many other countries of the MENA region. Available evidence on Se status among countries of the MENA region are summarized in Figure 2.

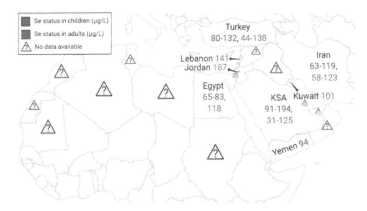

Figure 2. Available data on Se status in children and adults among countries of the MENA region.

4.1. Turkey

A study that analyzed Se concentrations in colostrum, transitional and mature breast milk among healthy lactating women concluded that Se content is below the international reference range set at 18.5 µg/l among all different samples throughout the lactation period [41]. Moreover, analysis was extended to dairy milk products in Turkey collected from different cities and its findings revealed that goat milk had the highest content followed by sheep milk and finally cow milk had the lowest Se content [41]. A more recent study included all dairy products in the analysis and found that concentrations varied based on the type of food with butter and cheese having higher concentrations; however, other products such as ice cream, milk, and yogurt had almost no Se content [42]. The estimated daily intake of Se among children was reported as 30-40 μg/day [43]. However, with regards to Se concentrations in blood, multiple case-control studies were conducted among different healthy and patient populations including infants, children, adults, pregnant women, and postmenopausal women. Other studies also used samples from erythrocytes and semen to assess Se concentrations [44,45]. Overall, values were widely variable ranging from 50–138 μg/l in healthy populations [43–55].

4.2. Jordan

A survey of groundwater was conducted to assess the levels of Se in multiple aquifers in Amman Zarqa Basin. Se levels were various among different aquifers ranging from 0.09 to 742 μg/l with an average of 24 μg/l, which exceeds WHO recommended threshold for drinking water [56]. A later case-control study of colorectal cancer patients assessed the dietary intake of Se which was found to be 59.26 ± 8.91 μg/day for controls. This value is higher than the RDA (55–70 μg/day) [57]. On the other hand, Se concentrations were analyzed in two studies using blood and hair samples [58,59]. Blood concentrations in non-smokers were found to be around 187 μg/l which is relatively high [58].

4.3. Iran

Multiple studies were conducted in Iran among various populations, children, adults, and elderly [60–62]. Se intake was found to be adequate in children and adults [60,61], whereas in postmenopausal women, Se intake was significantly lower than the RDA (55 μg/day) [62]. Se levels in rice samples from Iran were assessed and they were found to be relatively high [63]. Se status in Iran was assessed in various studies in healthy and patient populations. Blood concentrations ranged from 58–123 μg/l in children, adults and pregnant women [64–77].

4.4. Kingdom of Saudi Arabia (KSA)

A study was conducted to assess the content of Se in wheat grains grown in KSA. This study reported highly variable concentrations ranging from 8 to 293 μg/kg [78]. Other recent studies that quantified the intake of Se estimated the intake to be 75–121.65 μg/day and 93 μg/day in the regions of Jeddah and Riyadh, respectively [79,80]. Moreover, surveys of infant milk formulas, breast milk and cow's milk were conducted to determine Se intake in infant/children populations. The findings of those surveys showed that infant milk formulas contained adequate amounts of Se whereas some breast-fed infants might have lower than recommended Se intakes [81,82]. On the other hand, Se status was assessed in multiple case-control studies in KSA. Those studies were done using various samples; serum, urine, toenails, whole blood, umbilical cord blood, and placental tissue [5,80,83–91]. Blood concentrations varied widely from as low as 32 μg/l to 195 μg/l [5,80,85–88,90,91].

4.5. Egypt

Se intake among healthy children in Egypt was found to be 8.3 ± 2.3 mg/day [92]. Case-control studies in Egypt were mainly done on children, and reported Se concentrations ranging from 65–83 μg/l [93–95]. Studies have also assessed Se concentrations in neonates and their mothers and the concentrations were found to be 86 μg/l and 118 μg/l, respectively [96].

4.6. Qatar

In Qatar, no direct studies were conducted to assess the intake of Se among Qatari population. However, according to Qatar General Electricity & Water Corporation (Kahramaa) drinking water quality requirements report, Se is not expected to be present in Qatar's water system [97]. This indicates that the Qatari population is not expected to have any Se intake from drinking water sources.

Additionally, a market basket survey of Se in rice imports in Qatar, the primary staple dish, concluded that for Qatari citizens, rice compensates for more than 100% of RNI Se (30 µg/day) regardless of the gender or the type of rice consumed. However, for non-Qatari expatriates (over 80% of the population), with a much lower rice consumption, the percentages were variable according to gender and type of rice consumed, yet they were all below 100% of RNI Se [98]. Further analysis included rice-based infant cereals in Qatar which was found to provide around 63% of RNI Se based on the recommended daily serving [98]. This can conclude that rice consumption in Qatar is a significant contributor to the daily intake of Se. However, Se status in Qatar was not assessed in any of the few studies available.

5. Conclusions

Se plays a very important role in health and hence Se status and intake are very important to be known for a specific population as the levels of Se are highly variable among different populations and regions. Studies conducted in the MENA region to assess Se status and intake are very limited and most of those available were designed as case-control protocols with small sample sizes. For those countries with data available such as Iran, Turkey, and KSA, more powered, rigorous, and well-designed studies should be conducted to assess Se status among different populations in that specific country or regions that share more or less similar lifestyle and climate such as the six countries of the Gulf Cooperation Council (GCC). Se intake should also be assessed using food surveys and through surveying Se content in staple food (e.g., rice) especially that most of food products consumed in a large number of countries of the MENA region are imported from various parts of the world which may have variable Se content in the soil. Moreover, for many countries within the MENA region, including Qatar and most North-African countries, data about Se status is totally absent. This warrants the need for rigorous and well-constructed studies to determine Se intake and status among the general population and specific populations, such as children and pregnant women, in those countries. Furthermore, studying the relationship between Se status and the incidence of chronic diseases, such as diabetes and CVD in the MENA region would help in devising novel preventive approaches for these disorders particularly prevalent in many countries of this region.

Author Contributions: A.A. conceptualized the idea. A.A. and S.A.Z.I. wrote the manuscript. A.K. contributed to selected sections and edited the manuscript. A.A. coordinated the writing up and the submission process. All authors approved the final version for submission.

Funding: Our work is supported by award NPRP-8-1750-3-360 from Qatar National Research Fund (a member of Qatar Foundation) to A.A. The statements made herein are solely the responsibility of the authors.

Acknowledgments: The figures in this article were created with BioRender.

Conflicts of Interest: The authors declare no conflict of interest.

References

1. Weeks, M.E. The discovery of the elements. VI. Tellurium and selenium. *J. Chem. Educ.* **1932**, *9*, 474. [CrossRef]
2. Zachariah, M. High Selenium Induces Endothelial Dysfunction Via Endoplasmic Reticulum Stress. Ph.D. Thesis, University of Surrey, Guildford, UK, 2017.
3. Stoffaneller, R.; Morse, L.N. A Review of Dietary Selenium Intake and Selenium Status in Europe and the Middle East. *Nutrients* **2015**, *7*, 1494–1537. [CrossRef] [PubMed]
4. Prabhu, K.S.; Lei, X.G. Selenium. *Adv. Nutr.* **2016**, *7*, 415–417. [CrossRef] [PubMed]
5. Al-Daghri, N.M.; Al-Attas, O.; Yakout, S.; Aljohani, N.; Al-Fawaz, H.; Alokail, M.S. Dietary products consumption in relation to serum 25-hydroxyvitamin D and selenium level in Saudi children and adults. *Int. J. Clin. Exp. Med.* **2015**, *8*, 1305–1314. [PubMed]
6. Qazi, I.H.; Angel, C.; Yang, H.; Pan, B.; Zoidis, E.; Zeng, C.J.; Han, H.; Zhou, G.B. Selenium, Selenoproteins, and Female Reproduction: A Review. *Molecules* **2018**, *23*, 3053. [CrossRef] [PubMed]
7. Bleys, J.; Navas-Acien, A.; Guallar, E. Serum selenium levels and all-cause, cancer, and cardiovascular mortality among US adults. *Arch. Intern. Med.* **2008**, *168*, 404–410. [CrossRef]

8. Rocourt, C.R.; Cheng, W.H. Selenium supranutrition: Are the potential benefits of chemoprevention outweighed by the promotion of diabetes and insulin resistance? *Nutrients* **2013**, *5*, 1349–1365. [CrossRef]

9. Pieczyńska, J.; Grajeta, H. The role of selenium in human conception and pregnancy. *J. Trace Elem. Med. Biol.* **2015**, *29*, 31–38. [CrossRef]

10. Combs, F., Jr. Biomarkers of selenium status. *Nutrients* **2015**, *7*, 2209–2236. [CrossRef]

11. Papp, L.V.; Lu, J.; Holmgren, A.; Khanna, K.K. From selenium to selenoproteins: Synthesis, identity, and their role in human health. *Antioxid. Redox Signal.* **2007**, *9*, 775–806. [CrossRef]

12. Rayman, M.P. The importance of selenium to human health. *Lancet* **2000**, *356*, 233–241. [CrossRef]

13. Rayman, M.P. Selenium and human health. *Lancet* **2012**, *379*, 1256–1268. [CrossRef]

14. Xu, X.-M.; Carlson, B.A.; Mix, H.; Zhang, Y.; Saira, K.; Glass, R.S.; Berry, M.J.; Gladyshev, V.N.; Hatfield, D.L. Biosynthesis of selenocysteine on its tRNA in eukaryotes. *PLoS Biol.* **2007**, *5*, e4. [CrossRef] [PubMed]

15. Turanov, A.A.; Xu, X.-M.; Carlson, B.A.; Yoo, M.-H.; Gladyshev, V.N.; Hatfield, D.L. Biosynthesis of selenocysteine, the 21st amino acid in the genetic code, and a novel pathway for cysteine biosynthesis. *Adv. Nutr.* **2011**, *2*, 122–128. [CrossRef] [PubMed]

16. Schmidt, R.L.; Simonović, M. Synthesis and decoding of selenocysteine and human health. *Croat. Med. J.* **2012**, *53*, 535–550. [CrossRef] [PubMed]

17. Labunskyy, V.M.; Hatfield, D.L.; Gladyshev, V.N. Selenoproteins: Molecular pathways and physiological roles. *Physiol. Rev.* **2014**, *94*, 739–777. [CrossRef]

18. Kim, J.; Chung, H.S.; Choi, M.-K.; Roh, Y.K.; Yoo, H.J.; Park, J.H.; Kim, D.S.; Yu, J.M.; Moon, S. Association between Serum Selenium Level and the Presence of Diabetes Mellitus: A Meta-Analysis of Observational Studies. *Diabetes Metab. J.* **2019**, *43*, e1. [CrossRef] [PubMed]

19. Lewicka, I.; Kocyłowski, R.; Grzesiak, M.; Gaj, Z.; Oszukowski, P.; Suliburska, J. Selected trace elements concentrations in pregnancy and their possible role—Literature review. *Ginekol. Pol.* **2017**, *88*, 509–514. [CrossRef] [PubMed]

20. Mihailovic, M.; Cvetkovic, M.; Ljubic, A.; Kosanovic, M.; Nedeljkovic, S.; Jovanovic, I.; Pesut, O. Selenium and malondialdehyde content and glutathione peroxidase activity in maternal and umbilical cord blood and amniotic fluid. *Biol. Trace Elem. Res.* **2000**, *73*, 47–54. [CrossRef]

21. Mistry, H.D.; Broughton Pipkin, F.; Redman, C.W.; Poston, L. Selenium in reproductive health. *Am. J. Obstet. Gynecol.* **2012**, *206*, 21–30. [CrossRef] [PubMed]

22. Cengiz, B.; Soylemez, F.; Ozturk, E.; Cavdar, A.O. Serum zinc, selenium, copper, and lead levels in women with second-trimester induced abortion resulting from neural tube defects: A preliminary study. *Biol. Trace Elem. Res.* **2004**, *97*, 225–235. [CrossRef]

23. Zachara, B.A. Chapter Five-Selenium in Complicated Pregnancy. A Review. *Adv. Clin. Chem.* **2018**, *86*, 157–178.

24. Zhou, J.; Huang, K.; Lei, X.G. Selenium and diabetes-evidence from animal studies. *Free Radic. Biol. Med.* **2013**, *65*, 1548–1556. [CrossRef] [PubMed]

25. Clark, L.C.; Combs, G.F.; Turnbull, B.W.; Slate, E.H.; Chalker, D.K.; Chow, J.; Davis, L.S.; Glover, R.A.; Graham, G.F.; Gross, E.G.; et al. Effects of selenium supplementation for cancer prevention in patients with carcinoma of the skin. A randomized controlled trial. Nutritional Prevention of Cancer Study Group. *JAMA* **1996**, *276*, 1957–1963. [CrossRef] [PubMed]

26. Stranges, S.; Marshall, J.R.; Natarajan, R.; Donahue, R.P.; Trevisan, M.; Combs, G.F.; Cappuccio, F.P.; Ceriello, A.; Reid, M.E. Effects of long-term selenium supplementation on the incidence of type 2 diabetes: A randomized trial. *Ann. Intern. Med.* **2007**, *147*, 217–223. [CrossRef] [PubMed]

27. Lippman, S.M.; Klein, E.A.; Goodman, P.J.; Lucia, M.S.; Thompson, I.M.; Ford, L.G.; Parnes, H.L.; Minasian, L.M.; Gaziano, J.M.; Hartline, J.A.; et al. Effect of selenium and vitamin E on risk of prostate cancer and other cancers: The Selenium and Vitamin E Cancer Prevention Trial (SELECT). *JAMA* **2009**, *301*, 39–51. [CrossRef] [PubMed]

28. Thompson, P.A.; Ashbeck, E.L.; Roe, D.J.; Fales, L.; Buckmeier, J.; Wang, F.; Bhattacharyya, A.; Hsu, C.H.; Chow, H.H.; Ahnen, D.J.; et al. Selenium Supplementation for Prevention of Colorectal Adenomas and Risk of Associated Type 2 Diabetes. *J. Natl. Cancer Inst.* **2016**, *108*. [CrossRef]

29. Wang, X.-L.; Yang, T.-B.; Wei, J.; Lei, G.-H.; Zeng, C. Association between serum selenium level and type 2 diabetes mellitus: A non-linear dose-response meta-analysis of observational studies. *Nutr. J.* **2016**, *15*, 48. [CrossRef]

30. Benstoem, C.; Goetzenich, A.; Kraemer, S.; Borosch, S.; Manzanares, W.; Hardy, G.; Stoppe, C. Selenium and its supplementation in cardiovascular disease—What do we know? *Nutrients* **2015**, *7*, 3094–3118. [CrossRef]
31. Thomson, C.D. Assessment of requirements for selenium and adequacy of selenium status: A review. *Eur. J. Clin. Nutr.* **2004**, *58*, 391–402. [CrossRef]
32. Zhang, X.; Liu, C.; Guo, J.; Song, Y. Selenium status and cardiovascular diseases: Meta-analysis of prospective observational studies and randomized controlled trials. *Eur. J. Clin. Nutr.* **2016**, *70*, 162–169. [CrossRef] [PubMed]
33. Tan, H.W.; Mo, H.-Y.; Lau, A.T.Y.; Xu, Y.-M. Selenium Species: Current Status and Potentials in Cancer Prevention and Therapy. *Int. J. Mol. Sci.* **2018**, *20*, 75. [CrossRef] [PubMed]
34. Hatfield, D.L.; Tsuji, P.A.; Carlson, B.A.; Gladyshev, V.N. Selenium and selenocysteine: Roles in cancer, health, and development. *Trends Biochem. Sci.* **2014**, *39*, 112–120. [CrossRef] [PubMed]
35. Dreno, B.; Euvrard, S.; Frances, C.; Moyse, D.; Nandeuil, A. Effect of selenium intake on the prevention of cutaneous epithelial lesions in organ transplant recipients. *Eur. J. Dermatol. EJD* **2007**, *17*, 140–145. [CrossRef] [PubMed]
36. Karp, D.D.; Lee, S.J.; Keller, S.M.; Wright, G.S.; Aisner, S.; Belinsky, S.A.; Johnson, D.H.; Johnston, M.R.; Goodman, G.; Clamon, G.; et al. Randomized, double-blind, placebo-controlled, phase III chemoprevention trial of selenium supplementation in patients with resected stage I non-small-cell lung cancer: ECOG 5597. *J. Clin. Oncol.* **2013**, *31*, 4179–4187. [CrossRef] [PubMed]
37. Marshall, J.R.; Tangen, C.M.; Sakr, W.A.; Wood, D.P.; Berry, D.L.; Klein, E.A.; Lippman, S.M.; Parnes, H.L.; Alberts, D.S.; Jarrard, D.F.; et al. Phase III trial of selenium to prevent prostate cancer in men with high-grade prostatic intraepithelial neoplasia: SWOG S9917. *Cancer Prev. Res.* **2011**, *4*, 1761–1769. [CrossRef]
38. Algotar, A.M.; Stratton, M.S.; Ahmann, F.R.; Ranger-Moore, J.; Nagle, R.B.; Thompson, P.A.; Slate, E.; Hsu, C.H.; Dalkin, B.L.; Sindhwani, P.; et al. Phase 3 clinical trial investigating the effect of selenium supplementation in men at high-risk for prostate cancer. *Prostate* **2013**, *73*, 328–335. [CrossRef] [PubMed]
39. Lubinski, J.; Jaworska, K.; Durda, K.; Jakubowska, A.; Huzarski, T.; Byrski, T.; Stawicka, M.; Gronwald, J.; Górski, B.; Wasowicz, W.; et al. Selenium and the risk of cancer in BRCA1 carriers. *Hered. Cancer Clin. Pract.* **2011**, *9*, A5. [CrossRef]
40. Vinceti, M.; Filippini, T.; Del Giovane, C.; Dennert, G.; Zwahlen, M.; Brinkman, M.; Zeegers, M.P.; Horneber, M.; D'Amico, R.; Crespi, C.M. Selenium for preventing cancer. *Cochrane Database Syst. Rev.* **2018**, *1*, CD005195. [CrossRef]
41. Yanardag, R.; Orak, H. Selenium content of milk and milk products of Turkey. II. *Biol. Trace Elem. Res.* **1999**, *68*, 79–95. [CrossRef]
42. Ayar, A.; Sert, D.; Akin, N. The trace metal levels in milk and dairy products consumed in middle Anatolia-Turkey. *Environ. Monit. Assess.* **2009**, *152*, 1–12. [CrossRef] [PubMed]
43. Hincal, F. Trace elements in growth: Iodine and selenium status of Turkish children. *J. Trace Elem. Med. Biol.* **2007**, *21*, 40–43. [CrossRef]
44. Erkekoglu, P.; Asci, A.; Ceyhan, M.; Kizilgun, M.; Schweizer, U.; Atas, C.; Kara, A.; Kocer Giray, B. Selenium levels, selenoenzyme activities and oxidant/antioxidant parameters in H1N1-infected children. *Turk. J. Pediatr.* **2013**, *55*, 271–282. [PubMed]
45. Eroglu, M.; Sahin, S.; Durukan, B.; Ozakpinar, O.B.; Erdinc, N.; Turkgeldi, L.; Sofuoglu, K.; Karateke, A. Blood serum and seminal plasma selenium, total antioxidant capacity and coenzyme q10 levels in relation to semen parameters in men with idiopathic infertility. *Biol. Trace Elem. Res.* **2014**, *159*, 46–51. [CrossRef] [PubMed]
46. Ozdemir, H.S.; Karadas, F.; Pappas, A.C.; Cassey, P.; Oto, G.; Tuncer, O. The selenium levels of mothers and their neonates using hair, breast milk, meconium, and maternal and umbilical cord blood in Van Basin. *Biol. Trace Elem. Res.* **2008**, *122*, 206–215. [CrossRef]
47. Hincal, F.; Basaran, N.; Yetgin, S.; Gokmen, O. Selenium status in Turkey. II. Serum selenium concentration in healthy residents of different ages in Ankara. *J. Trace Elem. Electrol. Health Dis.* **1994**, *8*, 9–12.
48. Kilinc, M.; Guven, M.A.; Ezer, M.; Ertas, I.E.; Coskun, A. Evaluation of serum selenium levels in Turkish women with gestational diabetes mellitus, glucose intolerants, and normal controls. *Biol. Trace Elem. Res.* **2008**, *123*, 35–40. [CrossRef] [PubMed]
49. Seven, M.; Basaran, S.Y.; Cengiz, M.; Unal, S.; Yuksel, A. Deficiency of selenium and zinc as a causative factor for idiopathic intractable epilepsy. *Epilepsy Res.* **2013**, *104*, 35–39. [CrossRef]

50. Bay, A.; Dogan, M.; Bulan, K.; Kaba, S.; Demir, N.; Oner, A.F. A study on the effects of pica and iron-deficiency anemia on oxidative stress, antioxidant capacity and trace elements. *Hum. Exp. Toxicol.* **2013**, *32*, 895–903. [CrossRef]

51. Coskun, A.; Arikan, T.; Kilinc, M.; Arikan, D.C.; Ekerbicer, H.C. Plasma selenium levels in Turkish women with polycystic ovary syndrome. *Eur. J. Obstet. Gynecol. Reprod. Biol.* **2013**, *168*, 183–186. [CrossRef]

52. Arikan, D.C.; Coskun, A.; Ozer, A.; Kilinc, M.; Atalay, F.; Arikan, T. Plasma selenium, zinc, copper and lipid levels in postmenopausal Turkish women and their relation with osteoporosis. *Biol. Trace Elem. Res.* **2011**, *144*, 407–417. [CrossRef] [PubMed]

53. Sakiz, D.; Kaya, A.; Kulaksizoglu, M. Serum Selenium Levels in Euthyroid Nodular Thyroid Diseases. *Biol. Trace Elem. Res.* **2016**, *174*, 21–26. [CrossRef] [PubMed]

54. Arikan, T.A. Plasma Selenium Levels in First Trimester Pregnant Women with Hyperthyroidism and the Relationship with Thyroid Hormone Status. *Biol. Trace Elem. Res.* **2015**, *167*, 194–199. [CrossRef] [PubMed]

55. Aydemir, B.; Akdemir, R.; Vatan, M.B.; Cinemre, F.B.; Cinemre, H.; Kiziler, A.R.; Bahtiyar, N.; Buyukokuroglu, M.E.; Gurol, G.; Ogut, S. The Circulating Levels of Selenium, Zinc, Midkine, Some Inflammatory Cytokines, and Angiogenic Factors in Mitral Chordae Tendineae Rupture. *Biol. Trace Elem. Res.* **2015**, *167*, 179–186. [CrossRef] [PubMed]

56. Kuisi, M.A.; Abdel-Fattah, A. Groundwater vulnerability to selenium in semi-arid environments: Amman Zarqa Basin, Jordan. *Environ. Geochem. Health* **2010**, *32*, 107–128. [CrossRef] [PubMed]

57. Arafa, M.A.; Waly, M.I.; Jriesat, S.; Al Khafajei, A.; Sallam, S. Dietary and lifestyle characteristics of colorectal cancer in Jordan: A case-control study. *Asian Pac. J. Cancer Prev.* **2011**, *12*, 1931–1936. [PubMed]

58. Massadeh, A.; Gharibeh, A.; Omari, K.; Al-Momani, I.; Alomary, A.; Tumah, H.; Hayajneh, W. Simultaneous determination of Cd, Pb, Cu, Zn, and Se in human blood of jordanian smokers by ICP-OES. *Biol. Trace Elem. Res.* **2010**, *133*, 1–11. [CrossRef]

59. Alqhazo, M.; Rashaid, A.B. The concentrations of bioelements in the hair samples of Jordanian children who stutter. *Int. J. Pediatr. Otorhinolaryngol.* **2018**, *112*, 158–162. [CrossRef]

60. Mirzaeian, S.; Ghiasvand, R.; Sadeghian, F.; Sheikhi, M.; Khosravi, Z.S.; Askari, G.; Shiranian, A.; Yadegarfar, G. Assessing the micronutrient and macronutrient intakes in female students and comparing them with the set standard values. *J. Educ. Health Promot.* **2013**, *2*, 1. [CrossRef]

61. Darvishi, L.; Ghiasvand, R.; Ashrafi, M.; Ashrafzadeh, E.; Askari, G.; Shiranian, A.; Hasanzadeh, A. Relationship between junk foods intake and weight in 6–7 years old children, Shahin Shahr and Meymeh, Iran. *J. Educ. Health Promot.* **2013**, *2*, 2. [CrossRef]

62. Mansour, A.; Ahadi, Z.; Qorbani, M.; Hosseini, S. Association between dietary intake and seasonal variations in postmenopausal women. *J. Diabetes Metab. Disord.* **2014**, *13*, 52. [CrossRef] [PubMed]

63. Rahimzadeh-Barzoki, H.; Joshaghani, H.; Beirami, S.; Mansurian, M.; Semnani, S.; Roshandel, G. Selenium levels in rice samples from high and low risk areas for esophageal cancer. *Saudi Med. J.* **2014**, *35*, 617–620. [PubMed]

64. Safaralizadeh, R.; Kardar, G.A.; Pourpak, Z.; Moin, M.; Zare, A.; Teimourian, S. Serum concentration of selenium in healthy individuals living in Tehran. *Nutr. J.* **2005**, *4*, 32. [CrossRef] [PubMed]

65. Safaralizadeh, R.; Sirjani, M.; Pourpak, Z.; Kardar, G.; Teimourian, S.; Shams, S.; Namdar, Z.; Kazemnejad, A.; Moin, M. Serum selenium concentration in healthy children living in Tehran. *Biofactors* **2007**, *31*, 127–131. [CrossRef] [PubMed]

66. Dabbaghmanesh, M.H.; Sadegholvaad, A.; Ejtehadi, F.; Omrani, G. Low serum selenium concentration as a possible factor for persistent goiter in Iranian school children. *Biofactors* **2007**, *29*, 77–82. [CrossRef] [PubMed]

67. Farzin, L.; Moassesi, M.E.; Sajadi, F.; Amiri, M.; Shams, H. Serum levels of antioxidants (Zn, Cu, Se) in healthy volunteers living in Tehran. *Biol. Trace Elem. Res.* **2009**, *129*, 36–45. [CrossRef] [PubMed]

68. Farzin, L.; Sajadi, F. Comparison of serum trace element levels in patients with or without pre-eclampsia. *J. Res. Med. Sci.* **2012**, *17*, 938–941. [PubMed]

69. Farzin, L.; Moassesi, M.E. A comparison of serum selenium, zinc and copper level in visceral and cutaneous leishmaniasis. *J. Res. Med. Sci.* **2014**, *19*, 355–357.

70. Parizadeh, S.M.; Moohebati, M.; Ghafoori, F.; Ghayour-Mobarhan, M.; Kazemi-Bajestani, S.M.; Tavallaie, S.; Azimi-Nezhad, M.; Ferns, G.A. Serum selenium and glutathione peroxidase concentrations in Iranian patients with angiography-defined coronary artery disease. *Angiology* **2009**, *60*, 186–191. [CrossRef]

71. Rafraf, M.; Mahdavi, R.; Rashidi, M.R. Serum selenium levels in healthy women in Tabriz, Iran. *Food Nutr. Bull.* **2008**, *29*, 83–86. [CrossRef]

72. Tara, F.; Rayman, M.P.; Boskabadi, H.; Ghayour-Mobarhan, M.; Sahebkar, A.; Yazarlu, O.; Ouladan, S.; Tavallaie, S.; Azimi-Nezhad, M.; Shakeri, M.T.; et al. Selenium supplementation and premature (pre-labour) rupture of membranes: A randomised double-blind placebo-controlled trial. *J. Obstet. Gynaecol.* **2010**, *30*, 30–34. [CrossRef] [PubMed]

73. Mahyar, A.; Ayazi, P.; Fallahi, M.; Javadi, A. Correlation between serum selenium level and febrile seizures. *Pediatric Neurol.* **2010**, *43*, 331–334. [CrossRef] [PubMed]

74. Esalatmanesh, K.; Jamshidi, A.; Shahram, F.; Davatchi, F.; Masoud, S.A.; Soleimani, Z.; Salesi, M.; Ghaffarpasand, I. Study of the correlation of serum selenium level with Behcet's disease. *Int. J. Rheum. Dis.* **2011**, *14*, 375–378. [CrossRef] [PubMed]

75. Ghaemian, A.; Salehifar, E.; Shiraj, H.; Babaee, Z. A Comparison of Selenium Concentrations between Congestive Heart Failure Patients and Healthy Volunteers. *J. Tehran Heart Cent.* **2012**, *7*, 53–57. [PubMed]

76. Ghaemi, S.Z.; Forouhari, S.; Dabbaghmanesh, M.H.; Sayadi, M.; Bakhshayeshkaram, M.; Vaziri, F.; Tavana, Z. A prospective study of selenium concentration and risk of preeclampsia in pregnant Iranian women: A nested case-control study. *Biol. Trace Elem. Res.* **2013**, *152*, 174–179. [CrossRef] [PubMed]

77. Maleki, A.; Fard, M.K.; Zadeh, D.H.; Mamegani, M.A.; Abasaizadeh, S.; Mazloomzadeh, S. The relationship between plasma level of Se and preeclampsia. *Hypertens. Pregnancy* **2011**, *30*, 180–187. [CrossRef] [PubMed]

78. Al-Saleh, I.A.; Al-Jaloud, A.; Al-Doush, I.; El-Din, G. The distribution of selenium levels in Saudi dairy farms: A preliminary report from Al-Kharj. *J. Environ. Pathol. Toxicol. Oncol. Off. Organ Int. Soc. Environ. Toxicol. Cancer* **1999**, *18*, 37–46.

79. Al-Ahmary, K.M. Selenium content in selected foods from the Saudi Arabia market and estimation of the daily intake. *Arab. J. Chem.* **2009**, *2*, 95–99. [CrossRef]

80. Al-Othman, A.M.; Al-Othman, Z.A.; El-Desoky, G.E.; Aboul-Soud, M.A.; Habila, M.A.; Giesy, J.P. Daily intake of selenium and concentrations in blood of residents of Riyadh City, Saudi Arabia. *Environ. Geochem. Health* **2012**, *34*, 417–431. [CrossRef] [PubMed]

81. Al-Saleh, I.; Al-Doush, I. Selenium levels in infant milk formula. *Biometals Int. J. Role Met. Ions Biol. Biochem. Med.* **1997**, *10*, 299–302. [CrossRef]

82. Al-Saleh, I.; Al-Doush, I.; Faris, R. Selenium levels in breast milk and cow's milk: A preliminary report from Saudi Arabia. *J. Environ. Pathol. Toxicol. Oncol. Off. Organ Int. Soc. Environ. Toxicol. Cancer* **1997**, *16*, 41–46.

83. Al-Saleh, I.; El-Doush, I.; Billedo, G.; Mohamed Gel, D.; Yosef, G. Status of selenium, vitamin E, and vitamin A among Saudi adults: Potential links with common endemic diseases. *J. Environ. Pathol. Toxicol. Oncol. Off. Organ Int. Soc. Environ. Toxicol. Cancer* **2007**, *26*, 221–243. [CrossRef]

84. Al-Saleh, I.; Billedo, G. Determination of selenium concentration in serum and toenail as an indicator of selenium status. *Bull. Environ. Contam. Toxicol.* **2006**, *77*, 155–163. [CrossRef] [PubMed]

85. Al-Saleh, I.; Al-Doush, I.; Ibrahim, M.; Rabbah, A. Serum selenium levels in Saudi new-borns. *Int. J. Environ. Health Res.* **1998**, *8*, 269–275. [CrossRef]

86. Al-Saleh, I.; Billedo, G.; El-Doush, I.; El-Din Mohamed, G.; Yosef, G. Selenium and vitamins status in Saudi children. *Clin. Chim. Acta Int. J. Clin. Chem.* **2006**, *368*, 99–109. [CrossRef] [PubMed]

87. Al-Saleh, E.; Nandakumaran, M.; Al-Rashdan, I.; Al-Harmi, J.; Al-Shammari, M. Maternal-foetal status of copper, iron, molybdenum, selenium and zinc in obese gestational diabetic pregnancies. *Acta Diabetol.* **2007**, *44*, 106–113. [CrossRef] [PubMed]

88. Alissa, E.M.; Ahmed, W.H.; Al-ama, N.; Ferns, G.A. Selenium status and cardiovascular risk profile in healthy adult Saudi males. *Molecules* **2008**, *14*, 141–159. [CrossRef] [PubMed]

89. El-Yazigi, A.; Legayada, E. Urinary selenium in healthy and diabetic Saudi Arabians. *Biol. Trace Elem. Res.* **1996**, *52*, 55–63. [CrossRef] [PubMed]

90. Ghneim, H.K.; Alshebly, M.M. Biochemical Markers of Oxidative Stress in Saudi Women with Recurrent Miscarriage. *J. Korean Med. Sci.* **2016**, *31*, 98–105. [CrossRef]

91. El-Ansary, A.; Bjorklund, G.; Tinkov, A.A.; Skalny, A.V.; Al Dera, H. Relationship between selenium, lead, and mercury in red blood cells of Saudi autistic children. *Metab. Brain Dis.* **2017**, *32*, 1073–1080. [CrossRef]

92. Meguid, N.A.; Anwar, M.; Bjorklund, G.; Hashish, A.; Chirumbolo, S.; Hemimi, M.; Sultan, E. Dietary adequacy of Egyptian children with autism spectrum disorder compared to healthy developing children. *Metab. Brain Dis.* **2017**, *32*, 607–615. [CrossRef] [PubMed]

93. Saad, K.; Farghaly, H.S.; Badry, R.; Othman, H.A. Selenium and antioxidant levels decreased in blood of children with breath-holding spells. *J. Child Neurol.* **2014**, *29*, 1339–1343. [CrossRef] [PubMed]

94. Sherief, L.M.; Abd El-Salam, S.M.; Kamal, N.M.; El safy, O.; Almalky, M.A.A.; Azab, S.F.; Morsy, H.M.; Gharieb, A.F. Nutritional Biomarkers in Children and Adolescents with Beta-Thalassemia-Major: An Egyptian Center Experience. *BioMed Res. Int.* **2014**, *2014*, 1–7. [CrossRef] [PubMed]

95. Azab, S.F.A.; Saleh, S.H.; Elsaeed, W.F.; Elshafie, M.A.; Sherief, L.M.; Esh, A.M.H. Serum trace elements in obese Egyptian children: A case-control study. *Ital. J. Pediatr.* **2014**, *40*, 20. [CrossRef] [PubMed]

96. El-Mazary, A.-A.M.; Abdel-Aziz, R.A.; Mahmoud, R.A.; El-Said, M.A.; Mohammed, N.R. Correlations between maternal and neonatal serum selenium levels in full term neonates with hypoxic ischemic encephalopathy. *Ital. J. Pediatr.* **2015**, *41*, 83. [CrossRef] [PubMed]

97. Qatar General Electricity & Water Corporation (Kahramaa). *Overview on: Kahramaa Drinking Water Quality Requirement*; Kahramaa Publications: Doha, Qatar, 2014.

98. Rowell, C.; Kuiper, N.; Al-Saad, K.; Nriagu, J.; Shomar, B. A market basket survey of As, Zn and Se in rice imports in Qatar: Health implications. *Food Chem. Toxicol. Int. J. Publ. Br. Ind. Biol. Res. Assoc.* **2014**, *70*, 33–39. [CrossRef]

Review

Selenium, Selenoproteins and Viral Infection

Olivia M. Guillin [1,2,3,4,5], Caroline Vindry [1,2,3,4,5], Théophile Ohlmann [1,2,3,4,5] and
Laurent Chavatte [1,2,3,4,5,*]

[1]   CIRI, Centre International de Recherche en Infectiologie, CIRI, 69007 Lyon, France
[2]   Institut National de la Santé et de la Recherche Médicale (INSERM) Unité U1111, 69007 Lyon, France
[3]   Ecole Normale Supérieure de Lyon, 69007 Lyon, France
[4]   Université Claude Bernard Lyon 1 (UCBL1), 69622 Lyon, France
[5]   Centre National de la Recherche Scientifique (CNRS), Unité Mixte de Recherche 5308 (UMR5308),
      69007 Lyon, France
[*]   Correspondence: laurent.chavatte@ens-lyon.fr; Tel.: +33-4-72-72-86-24

Received: 30 July 2019; Accepted: 27 August 2019; Published: 4 September 2019

Abstract: Reactive oxygen species (ROS) are frequently produced during viral infections. Generation of these ROS can be both beneficial and detrimental for many cellular functions. When overwhelming the antioxidant defense system, the excess of ROS induces oxidative stress. Viral infections lead to diseases characterized by a broad spectrum of clinical symptoms, with oxidative stress being one of their hallmarks. In many cases, ROS can, in turn, enhance viral replication leading to an amplification loop. Another important parameter for viral replication and pathogenicity is the nutritional status of the host. Viral infection simultaneously increases the demand for micronutrients and causes their loss, which leads to a deficiency that can be compensated by micronutrient supplementation. Among the nutrients implicated in viral infection, selenium (Se) has an important role in antioxidant defense, redox signaling and redox homeostasis. Most of biological activities of selenium is performed through its incorporation as a rare amino acid selenocysteine in the essential family of selenoproteins. Selenium deficiency, which is the main regulator of selenoprotein expression, has been associated with the pathogenicity of several viruses. In addition, several selenoprotein members, including glutathione peroxidases (GPX), thioredoxin reductases (TXNRD) seemed important in different models of viral replication. Finally, the formal identification of viral selenoproteins in the genome of molluscum contagiosum and fowlpox viruses demonstrated the importance of selenoproteins in viral cycle.

Keywords: reactive oxygen species; glutathione peroxidases; thioredoxin reductases; influenza virus; hepatitis C virus; coxsackie virus; human immunodeficiency virus; molluscum contagiosum virus; viral selenoproteins; immunity

1. Introduction

Selenium is an essential trace element for mammalian redox biology. Numerous epidemiological studies have revealed an association between selenium deficiencies and the increased risks of developing several pathologies, including cancers, neurogenerative diseases, cardiovascular disorders and infectious diseases [1–13]. The ability of selenium supplementation to reverse or reduce these risks has been reported in many human or animal models although it remains controversial [14]. Unlike other trace elements that act as cofactors, selenium is covalently bound to organic molecules. Most of the beneficial effects of selenium is due to its incorporation in the form of selenocysteine into an essential group of proteins that are called selenoproteins. Selenocysteine is the 21st proteinogenic amino acid and is encoded by an UGA codon which is normally the signal for termination of protein synthesis [15–23]. Selenocysteine is a structural and functional analog of cysteine in which a selenium atom replaces sulfur to confer an enhanced catalytic activity. Amongst the twenty-five selenoprotein genes identified to date, several have important cellular functions in antioxidant defense, cell signaling and redox homeostasis [24]. Within the well characterized selenoproteins we find the following sub-families: Glutathione peroxidase (GPX1–GPX4 and GPX6) that reduce hydrogen and lipid peroxides [25], thioredoxin reductases (TXNRD1–TXNRD3) which are essentials in the homeostasis of thiol systems [26–29], methionine sulfoxide reductase (MSRB1) [30] and selenoproteins located in the endoplasmic reticulum (DIO2, SELENOF, SELENOK, SELENOM, SELENON, SELENOS and SELENOT) exhibit important functions in protein folding and in the endoplasmic reticulum stress response [31–33]. The other half of the selenoproteome remains without a, yet, defined function. Selenoproteins are present in many organelles or cellular compartments, with a specific tissue distribution and sensitivity to selenium level changes. Selenoproteins are therefore important components of antioxidant defense systems maintaining redox homeostasis, which also include catalase (CAT), superoxide dismutase (SOD), glutathione (GSH), vitamin E, carotenoids, and ascorbic acid.

Reactive oxygen species (ROS) are produced during viral infections with both beneficial and deleterious consequences for the cell (Figure 1). The viruses associated with ROS production are human immunodeficiency virus (HIV), hepatitis B virus (HBV), hepatitis C virus (HCV), Epstein-Barr virus (EBV), herpes simplex virus type 1 (HSV-1), vesicular stomatitis virus (VSV), respiratory syncytial virus (RSV), human T cell leukaemia virus type 1 (HTLV-1) and influenza viruses [34]. The mechanisms of ROS generation by the various viruses are diverse, but in several cases the host antioxidant defense enzymes, and especially members of the selenoproteome, are targeted.

2. Reactive Oxygen Species (ROS) in Immunity and Viral Infection

2.1. ROS and Oxidative Stress

The term "reactive oxygen species" (ROS) refers to series of side-products derived from molecular oxygen (O_2) generated during mitochondrial oxidative phosphorylation in every respiring cells (Figure 1). ROS can also arise from exogenous sources including drugs, xenobiotics, metals, radiation, smocking and infection [35]. ROS consist of radical and non-radical oxygen species formed by the partial reduction of molecular oxygen. They include superoxide anion radical ($O_2^{\bullet-}$), hydrogen peroxide (H_2O_2), and hydroxyl radical ($HO^{\bullet}$). At low concentration, ROS are also essential molecules in physiological processes such as cell signaling, proliferation, tumor suppression, and maintenance of the immune system. Oxidative stress arises when an imbalance between ROS and the cellular antioxidant defense system occurs (Figure 1). This could be due to an increase in ROS levels or a decrease in the cellular antioxidant capacity. Oxidative stress leads to direct or indirect ROS-mediated damage of nucleic acids, proteins, and lipids, and this phenomenon has been implicated in many pathological conditions including carcinogenesis [36], neurodegeneration [37,38], atherosclerosis, diabetes [39], and aging [40].

The production of ROS can be assessed indirectly either by using redox-sensitive dyes that are oxidized by ROS into quantifiable fluorescent products, such as 20,70-dichlorodihydrofluorescein diacetate (DCFHDA) or by quantification of cellular oxidation products such as oxidized DNA (8-hydroxydeoxyguanosine), lipids (malondialdehyde, F2-isoprostane, 7-ketocholesterol, and 7-hydroxycholesterol), proteins (carbonyl, 4-hydroxynonenal or glycated oxidation products). Many enzymatic assays are also available to evaluate the antioxidant function of the organisms [41].

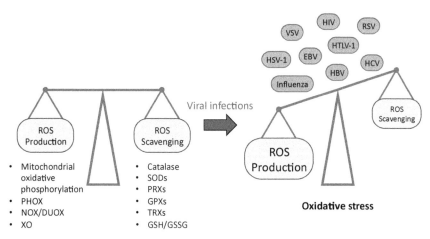

Figure 1. Balance between the generation of reactive oxygen species (ROS) and their scavenging systems in human. This equilibrium can be unbalanced during viral infections, resulting in oxidative stress. The main ROS producing systems include the mitochondrial oxidative phosphorylation, the phagocytic cell NAPDH oxidases (PHOX), the NADPH oxidases/dual oxidases (NOX/DUOX) and the xanthine oxidase (XO). The main ROS scavenging systems include the catalase, the superoxide dismutases (SODs), the peroxiredoxins (PRXs), the glutathione peroxidases (GPXs), the thioredoxins (TRXs) and the balance between reduced and oxidized glutathione (GSH/GSSG). The viruses for which an oxidative stress has been reported are herpes simplex virus type 1 (HSV-1), influenza viruses, vesicular stomatitis virus (VSV), Epstein-Barr virus (EBV), human immunodeficiency virus (HIV), human T cell leukaemia virus type 1 (HTLV-1), hepatitis B virus (HBV), respiratory syncytial virus (RSV) and hepatitis C virus (HCV) [34].

2.2. ROS Function in Immunity and Cell Signaling

ROS have an important role in host defense and immunity [42]. The most characterized example is the mechanism by which phagocytic cells produce large amounts of ROS to eliminate a wide variety of pathogens without altering the host cell viability. The nicotinamide adenine dinucleotide phosphate (NADPH) oxidase enzyme complex of phagocytic cells (PHOX) produces superoxide anion radical in the phagocytic vacuole via the transfer of one electron from NADPH to molecular oxygen [41]. The conventional idea is that this $(O_2^{\bullet-})$ molecule dismutates to form H_2O_2 and other ROS by further chemical or enzymatic reactions [43,44]. Indeed, the myeloperoxidase (MPO) that is an abundant protein released from the granules into the vacuole can further process H_2O_2 into HOCl. While the mechanism by which ROS can neutralize the invading micro-organisms in the phagosome is still a matter of debate, the production of HOCl by MPO seems to have a predominant role [43,44].

In addition to microbicidal activity, ROS also act as signaling mediators during cell death/apoptosis but also in processes that control cellular proliferation and differentiation. The family of NADPH oxidases (NOX) and Dual Oxidases (DUOX), referred to as NOX/DUOX, are homologs to PHOX and expressed in a variety of tissues, including colon, kidney, thyroid gland, testis, salivary glands, airways and lymphoid organs. A clear role for cytoplasmic ROS generated by NOX2 as well as DUOX1 has been shown in T cell receptor signaling as well as downstream activation and differentiation of T cells [45–48]. ROS production by mitochondrial complex III is required for antigen-induced T cell activation and production of interleukin-2 which is the cytokine essential for T cell proliferation [49].

2.3. ROS and Viral Infection

Viral infection is often accompanied by alteration of intracellular redox state of the host cell [34,41,50–60] (Figure 1). Viruses are known to induce ROS-generating enzymes, including NOX/

DUOX and xanthine oxidase (XO) and to disturb antioxidant defenses. XO is implicated in the catabolism of purine nucleic bases by producing H_2O_2. Increased of NOX/DUOX and XO activities were observed both in vitro and in vivo during viral infection [41]. Infection by the HIV is associated with decreased levels of GSH and an increased production of ROS [61–63]. The latter can be caused directly by virus and/or by the inflammatory response of the host. The viral TAT protein increases intracellular ROS levels by inhibiting the antioxidant enzyme manganese superoxide dismutase MnSOD [64]. In chronic hepatitis C, direct interaction of core protein with mitochondria is an important cause of the oxidative stress [57]. The increase of ROS production has been well documented during HIV, HBV, HCV, EBV, HSV-1, VSV, RSV, HTLV-1 and Influenza viral infections [34]. With HIV-1, ROS were found to stimulate viral replication with the nuclear transcription factor NF-kB, which is necessary for viral replication, being activated by oxidative stress in vitro [54,57,65].

3. Selenium, Selenoproteins and ROS

3.1. Selenium Insertion in Selenoproteins

Food is the primary source of selenium intake for mammals, but only five molecules (selenocysteine, selenomethionine, selenoneine, selenite, and selenate) constitute the bioavailable selenium in food intake [9,11]. The recommended daily intake of selenium in adults is comprised between 50 and 70 µg per day. A repeated daily intake above 400 µg leads to selenosis and eventually death. However, in certain regions of China, continual intakes of ~1000 µg Se/day are not associated with adverse effects other than fragile hair and fingernails due to keratin disruptions. Importantly, the concentration of selenium measured in the soil and water determines its levels in the living organisms and crops growing in these territories, and notably the components of human food chain, including microorganisms, plants, cereals, vegetables, fruits, farm animals, etc. The importance of selenium as a trace element in human health has been evidenced in a selenium-deprived area of China named Keshan, which then provided the name to the disease, as described in Section 4.1 and in [66]. Strikingly, Keshan disease has been fully eradicated by selenium supplementation [66]. Other regions around the globe are particularly deprived from selenium (<0.1 mg·kg^{-1}), and are located in China, New Zealand, Finland, South-East of USA, and in the UK [9].

It is now well admitted that the biological activity of selenium comes from its insertion into selenoproteins as a rare amino acid, selenocysteine [15–23]. The human selenoproteome is encoded by 25 selenoprotein genes and is highly regulated by selenium bioavailability [9,10,15,22,67–70]. Many reports have evidenced a prioritized regulation of the selenoproteome in response to selenium depletion that maintains the expression of essential selenoenzymes at the expense of others. Upon a normal diet, tissue concentration of selenium in the human body ranges from highest levels to lowest: Kidney, liver, spleen, pancreas, heart, brain, lung, bone and skeletal muscle [11]. Interestingly, in animals fed with a low selenium diet, selenium levels are drastically reduced in most tissues, including the normally selenium-rich ones kidneys and liver, but are maintained in very few tissues such as the brain and neuroencrine glands [71]. This phenomenon, described at the scale of organism, tissue, or cell lines, is referred to as selenium or selenoprotein hierarchy [15,22].

The process of selenocysteine insertion relies on a translational mechanism that is unique in many aspects. Selenocysteine was the first addition to the genetic code and is therefore referred to as the 21st amino acid. This amino acid is encoded by the UGA codon, which is normally a stop codon [15,17]. Thus, the cell has evolved a dedicated machinery to recode UGA as selenocysteine in selenoprotein mRNAs while maintaining its use as a stop codon in other cellular mRNAs [15,17]. The selenocysteine insertion sequence (SECIS) located in the 3' UTR of the mRNA [72] and the selenocysteine-tRNA (Sec-tRNA$^{[Ser]Sec}$) [16], together with their interacting protein partners allow the co-translational incorporation of a selenocysteine amino acid in selenoproteins. This mechanism is rather inefficient (between 1 and 5%), and mostly results in a truncated protein, the UGA codon being read as a stop codon [73–78]. Interestingly, more and more reports support the idea that the

UGA-selenocysteine recoding event by the ribosome is a limiting stage, and its efficiency dictates selenoprotein expression [15,17,22,67–70,74].

3.2. Role of Glutathione Peroxidases in Antioxidant Defense

Twenty five selenoprotein genes are present in the human genome [79–81], and most of them are involved in a redox reaction [24]. Among the selenoproteome, the GPXs are major components of the mammalian antioxidant defense. In humans, eight GPXs paralogs have been identified, five of them contain a selenocysteine residue in the catalytic site (GPX1–GPX4, GPX6), and three have a cysteine instead (GPX5, GPX7 and GPX8). Among the selenoenzymes, GPX1 and GPX4 are ubiquitously expressed and represent two of the most abundant selenoproteins in mammals. GPX1 is only cytoplasmic while GPX4 is localized in cytoplasmic, mitochondrial, and nuclear cellular compartments. GPX3 is a glycosylated protein secreted in the plasma mostly by the kidney, and its enzymatic activity is commonly used to evaluate the selenium status of the organism as its level fluctuates with selenium intake. GPX2 has initially been described as a gastrointestinal-specific enzyme but is present in other epithelial tissues (lung, skin, liver). Finally, the recently characterized GPX6, is only found in the olfactory epithelium and embryonic tissues. The role of the GPXs is to reduce hydrogen peroxides and organic hydroperoxides before they cause oxidative damage by reacting on cellular components. GPXs use GSH as a cofactor which is subsequently recycled by glutathione reductases [25], Figure 2A. In vitro, GPXs are able to reduce a wide variety of substrates that include H_2O_2, tert-butyl hydroperoxide, cumene hydroperoxide, ethyl hydroperoxide, linoleic acid hydroperoxide, paramenthane hydroperoxide, phosphatidylcholine hydroperoxide and cholesterol hydroperoxide (Figure 2A). As reported in [82,83] the different GPXs have overlapping enzymatic activities but they exhibit strong substrate specificities. For example, GPX4 is thought to be specialized in the reduction of lipid hydroperoxides while GPX1 is involved in the regulation of H_2O_2 metabolism.

The individual role of GPX members has been revealed by gene inactivation in mice. Interestingly, while the inactivation of *Gpx4* gene is embryonically lethal [84], mice deficient in *Gpx1* or *Gpx2* genes are perfectly healthy, fertile and show no increased oxidative stress as compared with wild-type (WT) animals in normal growth conditions [85–87]. However, when *Gpx1*$^{-/-}$ and WT mice are exposed to lethal doses of pro-oxidant, such as paraquat or H_2O_2, an eight-fold decrease in survival is observed for *Gpx1*$^{-/-}$ knockout (KO) mice [88,89]. This data suggests that GPX1 has a major role in protecting cells against strong oxidative stress but plays a limited role during normal development and under physiologic conditions. GPX2 appears to have a dual role in cancer, behaving either as a protector of carcinogenesis or a promoter of tumor growth, as revealed by various models [90].

3.3. Role of Thioredoxin Reductase in Antioxidant Defense, Redox Homeostasis and Redox Signaling

The two major reductive systems in mammalian cells are the thioredoxin (Txn) and GSH pathways. The Txn system is completely dependent on selenium as the three thioredoxin reductases (TXNRD1–TXNRD3) are selenoproteins with the selenocysteine residue at the penultimate position of the C-terminal end of the protein [26,28]. TXNRD1 and TXNRD2 are ubiquitously present in the cytoplasm and mitochondria, respectively, while TXNRD3 expression is restricted to specific tissues. The primary substrates of TXNRD1 and TXNRD2 are Txn1 and Txn2, respectively, Txn2 being localized in the mitochondria. TXNRDs catalyze the NADPH-dependent reduction of oxidized thioredoxin (Figure 2B). The Txns catalyze the reduction of protein disulfides such as in ribonucleotide reductase, peroxiredoxins (PRX), MSRB1, protein disulfide-isomerase (PDI), and are therefore critical for DNA synthesis, the defense against oxidative stress and disulfide formation within the endoplasmic reticulum [20]. Peroxiredoxins are able to reduce H_2O_2, organic hydroperoxides and peroxynitrite in order to protect cellular components from oxidative damage. However, the existence of multiple peroxide-removing enzymes such as catalase, GPX and PRX indicates that these peroxidases are not simply used in oxidant defense [91]. During inflammation, high levels of peroxides are produced by phagocytes to kill microorganisms. It has been well established that PRXs play cytoprotective antioxidant role in inflammation. Recently, it has been proposed that PRXs may play key roles in innate immunity and inflammation [91]. It becomes clear that, in addition to fighting oxidative stress, PRXs are important modulators of peroxide signaling. In addition to Txn, TXNRDs can reduce other small molecules containing sulfur, selenium, or oxidized semiquinone and therefore participate in many other cellular processes [20,26,28,92], Figure 2B.

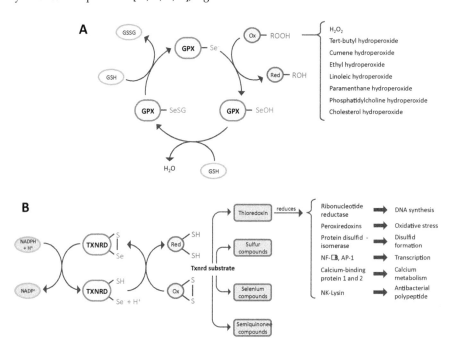

Figure 2. Enzymatic activities of the two most important families of selenoproteins involved in antioxidant defense in mammalian cells: the glutathione peroxidase (GPX) and the thioredoxin reductase (TXNRD). (**A**) GPXs use two molecules of reduced glutathione (GSH) to reduce hydrogen peroxides and organic hydroperoxides (ROOH) in their respective alcohols (ROH). The various peroxide substrates of mammalian GPXs are listed next to the bracket; (**B**) TXNRDs use NADPH to catalyze the reduction of thioredoxins and therefore participate in many cellular functions but can also reduce other sulfur or selenium containing compounds. Ox, oxidized molecule; Red, reduced molecule.

4. Selenium, Selenoproteins and Viral Replication

It is now recognized that the nutritional status of the host plays a leading role in the defense against infectious diseases [93–97]. Many studies show that nutritionally deficient humans or animals are more susceptible to a wide variety of infections. For a long time, researchers have believed that this was only the result of an impaired host immune response due to the deficiency of a particular nutritional element. However, as described below, the mechanism is more complex in that nutritional deficiency impacts not only the immune response of the host but also the viral pathogen itself. Thus, dietary selenium deficiency that causes oxidative stress in the host can alter a viral genome, so that a normally benign or mildly pathogenic virus becomes highly virulent in the deficient host under oxidative stress. This phenomenon has been reported in animal models for influenza and coxsackie viruses [95,98,99], but the molecular mechanism remains unclear. Once the viral mutations occur, even hosts with normal diet would be sensitive to the newly pathogenic strain. The link between selenium levels and viral infection has been reported for many viral groups, see Table 1.

4.1. Coxsackie Virus

The coxsackie virus is a nonenveloped, linear, positive-sense single-stranded RNA virus that belongs to the family of *Picornaviridae* (Group IV), genus *Enterovirus*. These enteroviruses, which also include poliovirus and echovirus, are among the most common and important human pathogens [100,101]. Coxsackie viruses are divided into group A (23 serotypes) and group B (six serotypes) viruses. In general, coxsackie viruses from group A infect the skin and mucous membranes, while viruses from group B infect the heart, pleura, pancreas, and liver [100].

In the early 1930s an endemic cardiomyopathy termed Keshan disease was first described in Heilongjiang province, Northeast China. This disease mainly affects infants, children and women in childbearing age [66]. It is characterized by cardiac enlargement, congestive heart failure, pulmonary edema and death. Keshan disease spread in another 12 provinces across China between the 1940s and 1960s. Approximately eight million people lived in the affected areas during that period of time, and thousands of people died of Keshan disease every year from this pathology. It is only in the 1970s and even the early 1980s, that the selenium contents in soil, water, food, and human body fluids were found extremely deficient in the areas affected by Keshan disease as compared with adjacent provinces [102]. Selenium fertilizer was applied to the soil in order to increase its content in the food [66]. In addition, selenium supplementation of the diet was also given to the people of these areas. The result was the complete eradication of this disease in these provinces of China [103]. However, several features of the Keshan disease, especially the annual or seasonal fluctuation in the incidence of the disease, did not wholly fit with a selenium deficiency. It appears that this disease has a dual etiology, i.e., selenium deficiency and an infectious cofactor, namely the coxsackie virus B [103–106].

Animal models were used to understand the relationship between host selenium nutritional status and coxsackie virus infection [53,93,94,96–99,107–112]. Coxsackie virus B3 (CVB3) infection of mice can cause myocarditis, similar to that found in human populations afflicted with Keshan disease. Interestingly, as illustrated in Figure 3 the work from Beck and co-authors showed that a non-virulent stain of CVB3 (designated CVB3/0) that do not lead to myocarditis, although replicating, is able to evolve in a virulent strain when inoculated in selenium deficient mice [98,99,109–111]. Remarkably, this is also true when *Gpx1* knockout mice were infected with the benign strain CVB3/0. The sequencing of the viral genomic RNA isolated from selenium deficient and *Gpx1*$^{-/-}$ mice demonstrated that a viral genome change had occurred during the infection and replication of the virus as compared to the viral genome replicated in selenium adequate animals, resulting in a highly pathogenic virus [108]. Out of the ten-nucleotide positions that were reported to co-vary with cardio-virulence in CVB3 strains, six reverted to the virulent genotype in virus isolated in Se-deficient mice, and seven in *Gpx1*$^{-/-}$ mice. Interestingly, a similar finding was also reported with the deficiency of another essential antioxidant, namely Vitamin E [93,94].

These experiments performed in animal models demonstrate that the host nutritional status, and particularly its antioxidant defense system is an important virulence factor, which can greatly contribute to the evolution of benign viral genomes into more virulent viruses. However, the molecular mechanism involved in this process remains to be elucidated.

4.2. Influenza Virus (Orthomyxoviridae)

Influenza viruses are enveloped, linear, negative-sense single-stranded RNA viruses belonging to the *Orthomyxoviridae* family (Group IV). There are four genus of this family: A, B, C and Thogotovirus, but only three influenza viruses are infectious for humans (A, B and C) [113]. The viral genome consists of eight segmented single-stranded RNA segments (seven for influenza C virus) encoding from 9 to 12 proteins, including hemagglutinin (HA) and neuraminidase (NA) surface glycoproteins, three ribonucleic acid (vRNA) polymerase subunits (vRNP: PA, PB1, PB2), non-structural protein (NS1), and matrix proteins M1 and M2 [113].

Various subtypes of the most common influenza A viruses are classified based on the diversity in the structure of HA and NA proteins. Influenza viruses can be divided into 16 different HA and

NA combinations. Influenza A and B viruses cause epidemics, whereas influenza C virus tends to cause infections with less severe symptoms [113]. According to the World Health Organization (WHO), the seasonal epidemics result every year in 3 to 5 million cases of severe illness and in 250 to 500 thousands deaths worldwide (https://www.who.int/influenza/en/). People at highest risk for mortality are the elderly and individuals with chronic diseases of the lung and heart. However, safe and effective vaccines are available but often do not perfectly match the circulating subtypes or become ineffective due to viral antigenic drift [113]. It is therefore necessary to engineer new vaccines and revaccinate people at risk every year.

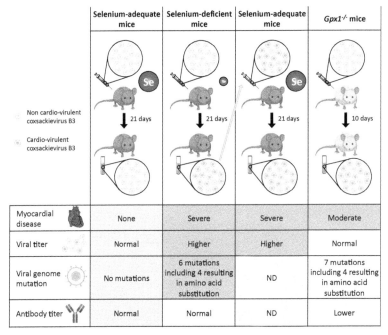

Figure 3. Evolution of the pathogenicity of Coxsackie virus as a function of selenium intake or selenoprotein knockout [53,93,94,96–99,107–111]. Coxsackie virus B3 (CVB3) infection of mice can cause myocarditis, similarly to that found in human disease. A non-virulent stain of CVB3 (referred to as CVB3/0, and shown in blue) does not lead to myocarditis in this animal model, although replicating in the mice heart fed with adequate selenium diet (left column). In case of selenium deficient mice, a group of animals was fed with a selenium-deficient diet for four weeks before infection with the benign strain CVB3/0 (second column from the left). A control group of animals was fed with an adequate-selenium diet and infected in parallel [98]. In case of selenium-deficient mice, they developed severe myocarditis. The sequencing of the CVB3 viral genome isolated from the heart of selenium-deficient mice showed mutations at nucleotide positions known to co-vary with cardio-virulence of CVB3 strains (shown in yellow). In comparison, the sequence of CVB3 isolated from selenium adequate mice showed no genetic variation (first column). To determine the consequences of the genetic alterations of the virus, CVB3 isolated from selenium deficient mice was inoculated in animals fed with a selenium-adequate diet (third column from the left) [98]. This experiment confirmed that the mutations of the viral genome increased the cardio-virulence of the virus, which can now induce severe myocarditis even in selenium adequate mice. To investigate whether the most abundant selenoprotein, GPX1, which expression correlates with selenium intake, is involved in the virulence of CVB3, a similar study was performed with *Gpx1*⁻/⁻ mice (right column) [108]. These mice, infected with the benign strain CVB3/0, developed myocarditis and nucleotide mutations of the viral genome isolated from their heart, similarly to selenium deficient mice.

The patients infected with influenza virus display a marked increase in DNA, lipid and protein oxidation products in blood plasma and urine [41,114–116]. Models of mice and cell lines infected with influenza viruses also show an enhanced production of ROS together with an imbalance of antioxidant defense [117–120]. These models are relevant to study the changes in redox homeostasis induced by the influenza virus.

The work from Beck's laboratory extended this novel concept that host nutritional status (especially selenium deficiency) is an important virulence factor in a viral family other than enteroviruses, as shown in Figure 4 [95,121–124]. Indeed, a rapid change in the pathogenicity of the virus in selenium deficient host has been also reported for influenza virus similarly to what was found for coxsackie virus. As shown in Figure 4, mice were fed with a diet either deficient or adequate in selenium for 4 weeks. Then, influenza A/Bangkok/1/79 (H3N2), a strain that induces mild pneumonitis in normal mice, was inoculated to both groups of mice. Interestingly, at all-time points post-infection a clear difference in pathology was observed between the two groups of mice [95,121–124]. The virus was much more virulent in selenium deficient mice, although with a similar virus titer than the selenium adequate mice. In addition, the sequencing of the HA, NA and M genes of viruses isolated from selenium-adequate and selenium-deficient mice demonstrated a strong impact of selenium status on virus mutation.

It appears that the selenium deficiency of the host promotes rapid genomic evolution of the virus in HA and NA genes as compared with selenium adequate animals [53,95,122–125]. Strikingly, these mutations are not stochastic as they were identical in three independent mice fed in selenium deficient diet. In comparison, very few mutations were detected in animals fed with adequate selenium diet. These data further confirm the impact of selenium status of the host in viral genome evolution.

4.3. Human Immunodeficiency Virus (HIV)

The human immunodeficiency virus (HIV) is an enveloped, linear, positive-sense single-stranded RNA virus that belongs to the family of *Retroviridae* (Group VI), genus *Lentivirus*. Two types of HIV have been characterized: HIV-1 and HIV-2 [126]. Given that HIV-1 is more virulent and more infective than HIV-2, HIV-1 has spread worldwide while HIV-2 is mostly confined to West Africa [127]. HIV is the etiologic agent of acquired immunodeficiency syndrome (AIDS) and is responsible for a weakened immune system as it infects immune cells [126]. HIV affects more than 35 million people worldwide and causes the death of about 1.5 million patients per year (http://www.who.int/hiv/en/). HIV infection is now considered as a chronic disease that requires intensive treatment and can present a variable clinical course. No vaccine is available until now, but an effective medication in decreasing the viral load and increasing the number of CD4 T-lymphocytes has been developed and is referred to highly active antiretroviral therapy (HAART) [128]. This treatment consists in the combination of three or more drugs that target different aspects of HIV replication [129].

HIV genome is highly compact and contains three genes encoding viral structural proteins (*gag, pol* and *env*), two genes for essential regulatory elements (*tat* and *rev*) and at least four genes encoding accessory regulatory proteins (*nef, vpr, vpu* and *vif*). As in any retrovirus, the RNA viral genome is reverse-transcribed in dsDNA that is then integrated in the host genome by the viral integrase. HIV-1 infects immune cells that harbor the CD4 receptor and a co-receptor belonging to the chemokine receptor family (CCR5 and CXCR4) [126]. Therefore, cells infected by HIV-1 are CD4 T-lymphocytes, monocytes, macrophages and dendritic cells. The replication but also the latency of the virus is extremely variable from one cell type to another.

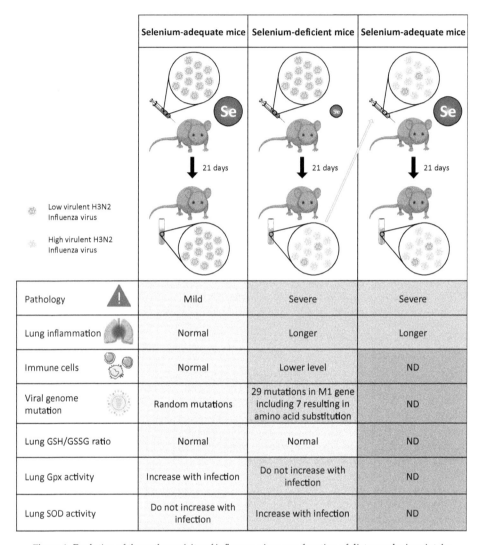

Figure 4. Evolution of the pathogenicity of influenza virus as a function of dietary selenium intake in mice. Influenza A/Bangkok/1/79 (H3N2) virus was inoculated in mice that were previously fed with selenium adequate or deficient diet for four weeks. This virus induces mild pneumonitis in selenium-adequate mice but a severe lung pathology in selenium deficient mice [95,121–124]. Various parameters, including the time of lung inflammation, the number of immune cells, the nucleotide mutations of the isolated influenza viruses, the oxidation status of glutathione (reduced/oxidized), the GPX and SOD enzymatic activities in the lung, were evaluated and compared between selenium adequate (left column) and deficient (middle column) mice. The low and high virulent H3N2 viruses are represented in blue and yellow respectively. The virus recovered from selenium-deficient mice was inoculated in selenium-adequate mice to evaluate its pathogenicity. Consistent with the observations made with coxsackie virus, the mutations of the influenza viral genome increased the pathogenicity of the virus, which can now induce severe lung pathology even in selenium adequate mice [95,121–124].

Lentiviruses are characterized by a long incubation period after the primo infection that is highly variable from one patient to another. During this time, humans infected with HIV are under chronic oxidative stress. The redox status of the patient is strongly disturbed in HIV infected patients as revealed by the decrease of antioxidant defense (selenium, ascorbic acid, alpha-tocopherol, carotenoids, superoxide dismutase, glutathione, and glutathione peroxidase) and the increase in ROS production (hydroperoxides, malondialdehyde, and clastogenic factors) [130]. The altered redox status seems to contribute to HIV pathogenesis in several ways. In vitro, increasing oxidative stress enhances the replication of HIV through the activation of NF-kB. Several mechanisms have been reported to explain the cellular enhancement of ROS production in HIV infection. Most of them imply the following viral proteins: Gp120, Tat, Nef, Vpr, and Retrotranscriptase (RT), as reviewed in [50]. A dramatic consequence of this chronic oxidative stress is the fatal decrease in the number of CD4 T-cells by apoptosis, and ultimately a failure of the immune system leading to death.

The nutritional deficiencies of the HIV-infected patient can affect the responsive capacity of the immune system and the progression to AIDS. Selenium is nowadays understood as an essential micronutrient for antioxidant defense and also immune function [131,132]. HIV infection simultaneously increases the demand for micronutrients and causes their loss which leads to a deficiency that can be compensated by micronutrient supplementation [133–135]. Low selenium levels are associated with a lower number of CD4 T-cells, faster progression of AIDS, and 20% increase in the risk of death [133,136]. However, little has been done in term of intervention studies by selenium supplementation or at the cellular and molecular levels to establish the link between selenium, selenoprotein and HIV infection. For example, selenium supplementation is only effective in slowing HIV progression for a subgroup of patients, for which serum selenium levels, CD4 count and viral load were improving in contrast to selenium non-responders or placebo group [137–139]. However, the cellular and molecular mechanism for this unequal response remains elusive. Although efficient at controlling viral load and restoring immune function, HIV antiretroviral therapies, especially the protease and reverse transcriptase inhibitors, have been shown to induce oxidative stress [50,140]. Interestingly, a long time treatment (more than 2 years) with antiretroviral therapy improves selenium levels as compared with HIV-infected patients not receiving the treatment [141].

The field awaits further investigations to understand the role of selenium and selenoproteins during HIV infection at the molecular level. The only in vitro data available reported a modification of the pattern of selenoprotein expression in response to HIV infection in lymphocytes [142] but these experiments were performed before the complete characterization of the selenoproteome. The impact of selenium status on viral genome mutations and in particular the shift to more virulent viruses has not yet been tested for HIV as it has been done for coxsackie and influenza.

4.4. Hepatitis C Virus (HCV)

The hepatitis C virus (HCV) is an enveloped, linear, positive-sense single-stranded RNA virus that belongs to the family of *Flaviviridae* (Groupe IV), genus *Hepacivirus*. Nowadays, about 3% of the world's population is infected with HCV, which represents approximately 170 million people. Although HCV replication occurs in hepatocytes, the virus also propagates in immune cells. In 80% of the patients with acute hepatitis C, the disease evolves to chronic hepatitis, with 2% developing liver cirrhosis and 1–5% developing liver cancer [143,144]. Many characteristics of oxidative stress have been reported during chronic hepatitis C, including a decrease in GSH, increase in MDA, HNE and caspase activity [145,146]. Zinc and selenium deficiencies increase the risk of chronicity and malignancy [147]. In addition, there is a high prevalence of HCV coinfection in HIV infected patients. The genome of around 9600 nucleotides encodes a unique polyprotein which is co- and post-translationally cleaved into 10 structural and non-structural proteins.

The infection by HCV is another well-documented example of virus-induced generation of ROS. The nucleocapsid protein of HCV, and to a lesser extent NS3, NS5A, E1, E2 and NS4B, are involved in generating oxidative stress in the liver [51,148–151]. In parallel, the plasma levels of selenium together with erythrocyte GPX activities were significantly lower in HCV-infected patients than in healthy controls. An inverse correlation of selenium levels with viral load was also observed [152]. Interestingly, in HCV and HIV co-infected patients, an even lower serum selenium concentration was measured than in HIV-infected patients [153]. Endoplasmic reticulum stress and unfolded protein response are induced by HCV gene expression [154]. A selenoprotein involved in these mechanisms, SELENOM, has been reported to be upregulated in human hepatocellular carcinoma (HCC) cell lines and liver biopsies of patients with HCV-related cirrhosis [155]. Whether this is true for other endoplasmic reticulum located selenoproteins remains to be investigated.

4.5. Other Viruses

The Hepatitis B virus is an enveloped virus with a circular and partially double-stranded DNA that belongs to the *Hepadnaviridae* family (Group VII). HBV includes several viruses that infect liver cells and cause hepatitis in humans and animals. In the viral genome, the large negative stranded DNA encodes the envelope, core and non-structural proteins, the DNA polymerase and an oncogenic transactivator [156,157]. The synthesis of the short strand is completed by cellular DNA polymerases after infection. There are 8 HBV strains, from A to H that differ from their geographic repartition [156,157]. Worldwide, between 2 and 8% of the population is infected by HBV but in most of the cases being asymptomatic. An acute HBV infection is however characterized by yellow eyes and skin, severe fatigue, vomiting and abdominal pain. In less than 5% of the cases, the infected people could develop a chronic infection which can further lead to a cirrhosis (in 20% of the cases) [156,157]. Several studies showed an association link between plasma selenium levels and progression of HBV infection [158–160]. For example, the selenium level is not correlated with the responsiveness to interferon treatment [161] but an elevated plasma selenium concentration is associated with a low level of transaminases [161]. Theses hepatic enzymes are implicated in amino acid catabolism, and their release in the plasma is linked to hepatocellular damage. In intervention studies, selenium supplementation decreased cancer incidence in HBV infected patients [162], but when the supplementation was stopped, the incidence became similar to control patients. Finally, in vitro, when hepatic cell lines were grown with different selenium concentration, lower viral proteins, viral transcripts and viral genomic DNA were detected with high selenium culture conditions [161].

The Porcine Circovirus 2 (PCV2) is a non-enveloped virus with a circular single-stranded DNA genome which belong to the *Circoviridae* family (Group II) [163]. Two strains exist, type 1 and type 2, but only type 2 causes a disease in swine, namely the Postweaning Multisystemic Wasting Syndrome (PMWS), a dramatic disease for pig-production industry. The severity of this syndrome is thought to highly dependent on intrinsic factor such as the status of the immune system. It is one of the smallest virus characterized so far, encoding only a capsid protein and two necessary proteins for viral replication [163]. It has been shown that selenomethionine supplementation in cell culture inhibits viral replication [164–168]. Furthermore, addition of H_2O_2 or ochratoxin A that induced oxidative stress enhanced viral replication. This effect was prevented by selenium supplementation [164–168] or by selenoproteins SELENOS and GPX1 [164–168]. It appears that this mechanism involved the autophagy pathway [164–168]. Finally, in infected mice, selenium supplementation was able to decrease histological lesions by reducing inflammation [164–168].

Table 1. Scientific literature available on the link between selenium, selenoprotein and viral infections listed as a function of Baltimore classification.

Group	Genome Structure	Virus Family	Virus	Epidemiological Study	Epidemiological Intervention	In Vitro Study	In Vivo Study	Viral Selenoprot
I	Double-stranded DNA	Herpesviridae	Epstein-Barr virus (EBV)	↓ GPX activity associated with ↑ viral load [169]		CT = ombilical blood mononuclear cells SS = Se-rich rice extract Inhibits EBV mediated cell transformation [170]		[171]
			Herpes Simplex Virus 2 (HSV-2)		SS = Selenium aspartate + multisupplementation Faster healing, ↓ in viral load and ↑ in antiviral cytokines [172]	CT = Vero cells SS = Diphenyl diselenide ↓ infectivity [173]		
			Human Herpesvirus 3 (HHV3)		SS = Selenium aspartate + multisupplementation Faster healing, ↓ in viral load and ↑ in antiviral cytokines [172]		AM = BALB/c Mice SS = Diphenyl diselenide ↓ histological damages and viral load ↑ levels of TNFalpha and IFNgamma [173]	
			Cytomegalovirus (CMV)					[171]
			Infectious bovine rhinotracheitis (IBRV)		SS = Sodium selenite ↑ GPX activity after infection in Se group ↑ IgM after infection in Se group ↑ antibody titer after infection in Se group [174]			
		Poxviridae	Molluscum contagiosum virus (MCV)					[171,175]
			Fowlpox virus (FWPV)					[176]
		Papovaviridae	Human Papillomavirus (HPV)		SS = Selenium aspartate + multisupplementation Faster healing, ↓ in viral load and ↑ in antiviral cytokines [172,177]			
II	Single-stranded DNA	Circoviridae	Porcine Circovirus type 2 (PCV2)			CT = PK15 cells SS = selenite, selenocarrageenan and selenomethionine Selenomethionine inhibits replication of PCV2 via Gpx1 and oxidative stress [164,165] CT = PK15 cells SELENOS overexpression may ↓ viral replication via oxidative stress [166] CT = PK15 cells SS = selenizing astragalus polysaccharide and selenomethionine ↓ PCV2 replication through autophagy ↓ [167,168]	AM = KunMing Mice SS = Selenized yeast ↓ TNFalpha, viral load and histological damages [128]	

Table 1. *Cont.*

Group	Genome Structure	Virus Family	Virus	Epidemiological Study	Epidemiological Intervention	In Vitro Study	In Vivo Study	Viral Selenoprot
IV	Positive-sense single-stranded RNA	*Picornaviridae*	Coxsackievirus B3 (CVB3)				AM = C3H/HeJ Mice SS = Selenite Apparition of a more virulent strain after infection of a selenium or vitamin E deficient host due to viral mutations [98,99,110,111,179] AM = Gpx1−/− Mice Apparition of a more virulent strain after infection of GPX1 deficient mice [108] AM = C57Bl/6 Mice SS = not specified Co-infection with a retrovirus lead to a more virulent pathology Selenium supplementation reverse the effect [180] AM = Balb/c Mice Viral load is associated with selenium status in tissues [181] AM = Mice SS = commercial Se repleted 'feedstuff' Se deficient mice exhibit a higher mortality, histopathological pathogenicity and viral load [182]	[171,183]
			Coxsackievirus B4 (CVB5)					[183]
			Coxsackievirus B5 (CVB5)			CT = Vero Cells SS = selenite, selenate and selenomethionine Selenite ↓ viral replication via thiol interaction [184]		
			Live attenuated poliomyelitis vaccine		SS = Sodium selenite ↑ GPX activity after infection in Se group ↑ antiviral cytokines ↑ TH4 response Faster clearance Mutations in the viral particles [185]			
			Foot-and-mouth disease virus (FMDV)		SS = selenium enriched yeast ↑ GPX activity after infection in Se group ↓ DNA damage [186]			

130

Table 1. *Cont.*

Group	Genome Structure	Virus Family	Virus	Epidemiological Study	Epidemiological Intervention	In Vitro Study	In Vivo Study	Viral Selenoprot
		Flaviviridae	Hepatite C virus (HCV)	↓ Se in infected people [152,160,187] ↓ GPX activity in infected people [152] No variation in GPX activity with infection [145]	SS = Selenomethionine ↑ GPX activity after infection in Se group No effect on viral load [188]			[183,189]
			West nile virus (WNV)			CT = Vero cells Selenium deficiency induces higher cell death and cytopathic effects but has no impacts on viral production [190]		
			Japanese encephalitis virus (JEV)					[191]
		Bunyaviridae	Hantaan virus (HTNV) or Seoul virus (SEOV)	↑ incidence of the infection with ↓ Se [192]		CT = HUVEC SS = sodium selenite ↓ viral replication with a low MOI [192]		
			Respiratory syncytial virus (RSV)	↓ Se in infected people [193]	SS = Sodium selenite Faster healing in Se group [194]			
V	Negative-sense single-stranded RNA	Filoviridae	Ebola virus (EBOV)					[195]
		Orthomyxoviridae	Influenza A/Bangkok/1/79 (H3N2)			CT = Differentiated human bronchial epithelial cells Se deficiency ↑ mucus production, influenza-induced apoptosis and modifies cytokine expression [122]	AM = C57Bl/6J SS = Selenite Deficiency leads to an ↑ in inflammation and pathology and an altered cytokine expression No changes in viral load [95] Apparition of a more virulent strain after infection of a selenium deficient host due to mutations [121] ↑ SOD activity in selenium deficient group [124]	
			Influenza A/Puerto Rico/8/34 (H3N2)	↓ Se in infected people [116,196]			AM = transgenic mice carrying a mutant Sec-tRNA[Ser]Sec No impact on lung pathology [123]	
			Influenza A (H1N1)	↓ GPX activity in infected people [116]			AM = C57Bl/6J SS = Selenite Deficiency leads to altered immune response responsible of less death No change in viral load [125] AM = KunMing Mice SS = Selenite Reduces mortality, ↑ levels of TNFalpha and IFNgamma No change in viral load [197]	

Table 1. *Cont.*

Group	Genome Structure	Virus Family	Virus	Epidemiological Study	Epidemiological Intervention	In Vitro Study	In Vivo Study	Viral Selenoprot
V	Negative-sense single-stranded RNA	*Orthomyxoviridae*	Avian influenza (H9N2)		SS sodium enriched yeast or sodium selenite ↓ viral shedding ↑ ISG expression and IFN [198]			
			Avian Influenza A/duck/Novosibirsk56/05 (H5N1)			CT = RK, BHK21 and Vero E6 cells SS = nutrient mixture containing selenium ↓ viral replication in late stages [199]		
			Parainfluenza-3 (PI3)		SS = sodium selenite ↑ GPX activity after infection in Se group ↑ IgM after infection in Se group antibody titer after infection in Se group [200]			
		Paramyxoviridae	Human metapneumovirus (HMPV)	↓ Se in infected people [193]				
			Measles virus (MV)					[183]
VI	Single-stranded RNA with a DNA intermediate	*Retroviridae*	Human immuno-deficiency virus 1 (HIV-1)	↓ Se in infected people [153,187,201–212]	No change in viral load SS = selenized yeast [137,139] SS = selenomethionine [219] SS = Sodium selenite [220] SS = not indicated [138,221]	CT = Jurkat and HeLa cells Infection or TAT expression ↓ some selenoproteins but ↑ low molecular mass selenocompounds [142,226]	In patients, a polymorphism a SELENOF is associated with a shorter time of progression to AIDS [232]	[195,233,234]
				No significant variation of Se level in infected people [213–215]	↑ in CD4 count in Se group SS = selenized yeast [137,139] SS = Sodium selenite [220] SS = not indicated [138,221]	CT = ACH2, Jurkat, ESb-L, KK1, U1 cells and monocytes SS = selenite Prevents HIV transcription by TNF alpha mediated NFkappaB activiation in chronically infected cells [227–229]		
				Low selenium level associated with low CD4 count [153,203,206,207,211]	No change in CD4 count SS = selenomethionine [219,222] SS = not indicated [223]	CT = SupT1 GPX1 overexpression ↑ viral replication and cytopathic effects and inversely [230]		
				Low selenium level associated with a higher progression to AIDS [201–204,206]	↑ of viral shedding in Se group SS = selenomethionine [224] SS = not indicated [221]	CT = U937, monocytes derived macrophages TXNRD1 negatively regulates TAT activity by targeting disulfides bonds [231]		
				Low selenium level associated with vaginal shedding of HIV [216]	Se supplementation improves child survival if the mother is infected SS = selenomethionine [219]			
				High selenium level associated with vaginal shedding of HIV [217]	Se supplementation decreases diarrheal morbidity SS = selenomethionine [225]			
				No significant variation of Se level in infected people treated with HAART [218]				
				More skin desease in Se deficient HIV infected people [211]				

Table 1. Cont.

Group	Genome Structure	Virus Family	Virus	Epidemiological Study	Epidemiological Intervention	In Vitro Study	In Vivo Study	Viral Selenoprot
VI	Single-stranded RNA with a DNA intermediate	*Retroviridae*	Human immuno-deficiency virus 2 (HIV-2)					[183]
			Simian immuno-deficiency virus (SIV)	↓ Se in infected monkeys [23]		CT = CEM and Jurkat cells Infection leads to a ↓ in selenoprotein expression and an ↑ in low molecular mass selenocompounds TAT transfection leads to a ↓ in GPX and SELENOF but an ↑ in TXNRD1 expression [235]		
			Murine Leukemia virus (MuLV)		SS = Sodium selenite ↑ GPX activity after infection in Se group [236]			[171]
VII	Double-stranded DNA with a single stranded RNA intermediate	*Hepadnaviridae*	Hepatitis B (HBV)	↓ Se in infected people [158,160] {↑ Se associated with less hepatic damages Abediankenari, 2011 #4334}	SS = selenized table salt or selenized yeast Lower cancer induced by HBV incidence [162]	CT = HepG2 and HuH7 SS = sodium selenite Suppresses HBV replication, transcription and protein expression [161]		[171]

SS, type of selenium supplementation used in the study; CT, cell type used for the study; AM, animal model used for the study; Se, selenium; ↓ decrease; ↑ increase.

5. Selenoproteins in Viral Genomes

5.1. 1998: First Example of a Viral Selenoprotein Encoded in Molluscum Contagiosum Virus Genome

Molluscum contagiosum is a viral infection that affects the skin and is caused by the dermatotropic poxvirus molluscum contagiosum virus (MCV) [237,238]. Unlike smallpox and human monkeypox diseases, MCV is nonlethal, mostly common in children and young adults and present worldwide [237,238]. However, MCV causes severe skin infections in immunosuppressed adults [237,238]. A typical feature is the apparition of single or multiple papules on the skin, which may persist for a few years. Most cases resolve in six to nine months without specific treatments [237,238]. Such a prolonged infection implies that MCV successfully manipulates the host environment. In 1998, the analysis of the MCV genome sequence predicted the presence of a candidate selenoprotein, homologous to mammalian GPX, with 75% amino acid sequence identity with human GPX1 [175], see Figure 5A,B. This viral GPX protein is encoded by *MC066L* gene that presents every features of a selenoprotein gene, i.e., an in-frame UGA codon, a stop codon different from an UGA (in this case UAG), and a SECIS element in the 3'UTR of the mRNA (Figure 5A). The absence of homologs of this gene in vaccinia and variola viruses suggests that the GPX-like gene was acquired by the MCV after the divergence of the *Molluscipoxvirus* and *Orthopoxvirus* genera. The expression of this predicted selenoprotein was tested experimentally in mammalian cells. Indeed, when a plasmid containing the *MC066L* gene was transfected in human skin cell lines, many evidences supported the insertion of a selenocysteine residue at the UGA codon in the full-length protein, the functionality of the SECIS elements and the cellular antioxidant activity of the MC066L protein [175,239]. Remarkably, this viral selenoprotein has been shown to be protective for human keratinocytes against cytotoxic effects of UV-irradiation and hydrogen peroxides [175,239], suggesting an important function for the virus in defending itself against environmental stress and inflammation. How and when this selenoprotein is expressed in the context of viral infection remains poorly investigated. The first transcription map of the MCV genome was provided by the transcriptome sequencing (RNA-seq) of the RNAs synthesized in abortively infected cultured cells and human skin lesions [240]. These next generation sequencing experiments showed that MC066L mRNA was only detected in cutaneous lesions, but not in MRC-5, Huh7.5.1 and Vero cells infected in vitro by the MCV virus isolated from these same skin lesions.

5.2. 2007: A Second Example of an Encoded Viral Selenoprotein in Fowlpox Virus Genome

Almost ten years later, another example of an encoded viral selenoprotein was reported in fowlpox viral genome [176], see Figure 5C,D. This was due to the increasing number of viral genome sequenced but also to the development of novel bioinformatic tools dedicated to the discovery of selenoprotein genes in newly sequenced genomes [79–81]. In this viral genome, a coding region homolog to the mammalian GPX4 gene was found, with an in-frame UGA codon, and a predicted SECIS element downstream of the UGA codon but this time within the open reading frame instead of being in the 3'UTR. This finding represented a great opportunity to investigate whether this putative viral SECIS or a canonical SECIS could function within the open reading frame. The authors demonstrated that mammalian cell lines supported the expression of selenoproteins with in-frame SECIS element from both viral and mammalian origin. This fowlpox SECIS element was the second example of a functional viral SECIS element with a structure being identical to the mammalian SECIS. Interestingly, in an evolutionary related virus, the canarypox virus (CPV), this gene has evolved in a Cys-containing GPX4 with a fossil SECIS element still present in the coding region (Figure 5C,D). The potential of this fossil SECIS to trigger recoding of an UGA codon in selenocysteine has not been investigated. It appears that there was a recent mutation of the selenocysteine into cysteine codon in canarypox virus, as it has happened multiple times during evolution of the selenoproteomes in *Eukarya*, *Bacteria* and *Archaea* [79,80]. Note that cysteine is encoded by UGC and UGU codons, and that a single mutation is able to change a selenocysteine to cysteine codon and vice-versa. The presence of a fossil SECIS element indicates that the GPX4 selenoprotein gene was first acquired from the host and recently converted to the Cys form.

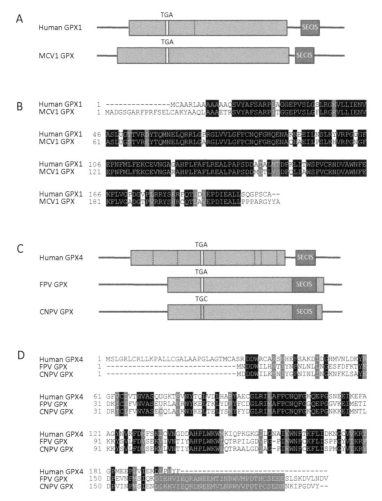

Figure 5. Gene structures and amino acid sequences of the selenoproteins present in the viral genomes of molluscum contagiosum virus subtype 1 (MCV1) (**A,B**) and of fowlpox virus (FPV) (**C,D**) in comparison with their respective human orthologs, GPX1 and GPX4 [175,176]. (**A**) Location of the typical features of a selenoprotein gene (coding sequence, TGA codon and SECIS element) in human *GPX1* gene in comparison with those of MCV1 *GPX* gene. For clarity reasons, the introns of human gene have been removed, but the position of splice sites is indicated by a dashed bar. (**B**) Amino acid sequence alignment of human GPX1 (P07203) with MCV1 GPX (Q98234). Identical and similar amino acids in both sequences are highlighted in black and grey, respectively. The selenocysteine amino acid (U, in one-letter code) is highlighted in yellow. (**C**) Comparison of the location of selenoprotein gene features in human *GPX4* gene with those of FPV *GPX* and Canarypox virus (CPV) *GPX* genes. The replacement of the TGA (selenocysteine) codon by a TGC (Cysteine) one in CPV *GPX* gene is indicated in green. (**D**) Amino acid sequence alignment of human GPX4 (P36969) with FPV GPX (Q70H87) and CPV GPX (Q6VZR0). In-frame SECIS elements in the C-terminal region of FPV GPX are highlighted in red.

The fact that at least two selenoproteins are encoded by viral genomes suggests that these proteins provide a substantial advantage for viruses. Similar to molluscum congatiosum GPX1, the fowlpox GPX4 may provide survival benefits for the virus. These two proteins are, so far, the only proven examples of genetically encoded viral selenoproteins.

5.3. Putative Selenoproteins in Other Viral Genomes

These two examples of selenoprotein gene snatching from eukaryotic genomes in the viral genomes of fowlpox and molluscum contagiosum viruses lead to the careful investigation for further examples of selenoprotein genes sequences with viral genomes. The first analysis searched for GPX modules within viral genomes where an in-frame UGA codon would be in an amino acid environment close to the catalytic site sequence of eukaryotic GPXs. Several candidates with sequence identities greater than 25% were found in the genomes of HIV-1, HIV-2, HCV, coxsackie virus B3 and measles viruses [183]. Despite these in silico data that GPX-related features are present in a number of RNA viruses, no RNA structure similar to the SECIS element can be evidenced. Additionally, no biochemical data demonstrated the expression of viral selenoproteins in any of these cases. It is possible that viruses have developed somewhat unique mechanisms for Sec insertion, as suggested in [195], but this remains purely hypothetical in the absence of further experimental proofs.

Perhaps, the most advanced study concerns a putative GPX protein coded in the third reading frame of the envelope (Env) gene of HIV-1 [241]. Indeed, it contains the typical catalytic triad selenocysteine (U), Glutamine (Q) and Tryptophan (W) and this putative HIV-GPX protein has been predicted to adopt the overall GPX fold, as deduced from computerized calculations [233]. In addition, it appears that the HIV-GPX gene is conserved in laboratory strains of HIV-1, as well as in long-term non-progressor isolates, but most of HIV isolates from patients with progressive disease presented deleterious mutations (mostly premature stop codons). In order to grasp the cellular function of this putative HIV-GPX in mammalian cells, the corresponding coding sequence has been fused to a mammalian SECIS element and transfected in mammalian cells [234]. The expression of the HIV-GPX seems to have an anti-apoptotic activity, by conferring cytoprotection against exogenous or endogenous ROS. Indeed, several viral proteins are known to induce apoptosis via redox-sensitive effects during HIV-1 viral cycle. Therefore, the presence of a HIV-GPX could be pertinent in the long-term non-progressor patients. Note that these experiments were performed before the emergence of the HAART.

Another putative viral selenoprotein gene has been reported in the -1 reading frame of the NS4 region of Japanese encephalitis virus (JEV). JEV belongs to the *Flaviviridae* family, which also includes dengue fever virus (DENV), yellow fever virus (YFV) and West Nile virus (WNV). The gene named NS4-fs encodes a potential 104 amino acid sequence with three predicted selenocysteine residues, i.e., three in frame UGA codons [191]. This putative selenoprotein displays 30.3% identity and 45.8% similarity with an aligned family of ferredoxin with cysteine instead of selenocysteine. Noteworthy, these three UGA codons are highly conserved, as they are present in all of the 15 full genomic JEV sequences analyzed. A 3D structure of the protein has been modeled [191] where the selenocysteine residues are proposed to maintain the conformation of the $[Fe_2S_2]$ cluster center. Interestingly, ferredoxin usually acts as an electron transfer agent in biological redox reactions, and this may somehow be important for JEV infection or replication. Again, in this example, neither SECIS elements were found nor any biochemical evidence of selenoprotein expression was provided.

6. Conclusions

During viral infections, there are many ways that the host metabolism could be affected, leading to a dysregulation of redox homeostasis. The viral pathogens induce oxidative stress via the increase generation of ROS and the alteration of cellular ROS scavenging systems. As part of antioxidant defense, selenoproteins, such as GPXs, TXNRDs and those located in the ER, play an important role in controlling oxidative stress. Selenium deficiency creates a weakening of the defense against infectious diseases by reducing selenoprotein expression. However, nutritional status of the host can also lead to viral genome mutations from a benign or mildly pathogenic virus to a highly virulent one under oxidative stress that could further spread in hosts with adequate selenium intake. The molecular mechanism leading to the site-specific genome evolution of the virus toward more pathogenic strains awaits further experiments, especially to understand the implication of the selenoproteins.

Funding: This work was supported by ANRS, INSERM, CNRS and ENS de Lyon 'Emerging Project' (LC). OMG was a recipient of a Master fellowship from Labex Ecofect (ANR-11-LABX-0048) of the Université de Lyon, within the program Investissements d'Avenir (ANR-11-IDEX-0007) operated by the French National Research Agency (ANR). OMG is a recipient of a Ph.D fellowship from the Université Claude Bernard Lyon 1. CV is a recipient of a Post-doctoral fellowship from ANRS.

Acknowledgments: In several figures and in the graphical abstract, we used illustrations from (https://smart. servier.com/) and (https://biorender.com/) websites.

Conflicts of Interest: The authors declare no conflict of interest.

References

1. Rayman, M.P. Selenium and human health. *Lancet* **2012**, *379*, 1256–1268. [CrossRef]
2. Rayman, M.P. The importance of selenium to human health. *Lancet* **2000**, *356*, 233–241. [CrossRef]
3. Hatfield, D.L.; Tsuji, P.A.; Carlson, B.A.; Gladyshev, V.N. Selenium and selenocysteine: Roles in cancer, health, and development. *Trends Biochem. Sci.* **2014**, *39*, 112–120. [CrossRef] [PubMed]
4. Hatfield, D.L.; Carlson, B.A.; Xu, X.M.; Mix, H.; Gladyshev, V.N. Selenocysteine incorporation machinery and the role of selenoproteins in development and health. *Prog. Nucleic Acid Res. Mol. Biol.* **2006**, *81*, 97–142. [PubMed]
5. Meplan, C.; Hesketh, J. Selenium and cancer: A story that should not be forgotten-insights from genomics. *Cancer Treat. Res.* **2014**, *159*, 145–166. [CrossRef] [PubMed]
6. Papp, L.V.; Holmgren, A.; Khanna, K.K. Selenium and selenoproteins in health and disease. *Antioxid Redox Signal* **2010**, *12*, 793–795. [CrossRef] [PubMed]
7. Whanger, P.D. Selenium and its relationship to cancer: An update. *Br. J. Nutr.* **2004**, *91*, 11–28. [CrossRef]
8. Kurokawa, S.; Berry, M.J. Selenium. Role of the essential metalloid in health. *Met. Ions Life Sci.* **2013**, *13*, 499–534. [CrossRef]
9. Vindry, C.; Ohlmann, T.; Chavatte, L. Selenium metabolism, regulation, and sex differences in mammals. In *Selenium, Molecular and Integartive Toxicology*; Michalke, B., Ed.; Springer International Publishing AG, Part of Springer Nature: Cham, Switzerland, 2018; pp. 89–107.
10. Touat-Hamici, Z.; Legrain, Y.; Sonet, J.; Bulteau, A.-L.; Chavatte, L. Alteration of selenoprotein expression during stress and in aging. In *Selenium: Its Molecular Biology and Role in Human Health*, 4th ed.; Hatfield, D.L., Su, D., Tsuji, P.A., Gladyshev, V.N., Eds.; Springer Science+Business Media, LLC: New York, NY, USA, 2016; pp. 539–551.
11. Sonet, J.; Bulteau, A.-L.; Chavatte, L. Selenium and Selenoproteins in Human Health and Diseases. In *Metallomics: Analytical Techniques and Speciation Methods*; Michalke, B., Ed.; Wiley-VCH Verlag GmbH & Co. KGaA: Weinheim, Germany, 2016; pp. 364–381. [CrossRef]
12. Latrèche, L.; Chavatte, L. Selenium incorporation into selenoproteins, implications in human health. *Met. Ions Biol. Med. X* **2008**, *10*, 731–737.
13. Avery, J.C.; Hoffmann, P.R. Selenium, Selenoproteins, and Immunity. *Nutrients* **2018**, *10*, 1203. [CrossRef]
14. Jackson, M.I.; Combs, G.F. Selenium as a Cancer Preventive Agent. In *Selenium: Its Molecular Biology and Role in Human Health*, 3rd ed.; Hatfield, D.L., Berry, M.J., Gladyshev, V.N., Eds.; Springer: New York, NY, USA, 2012; pp. 313–323.
15. Vindry, C.; Ohlmann, T.; Chavatte, L. Translation regulation of mammalian selenoproteins. *Biochim. Biophys. Acta Gen. Subj.* **2018**, *1862*, 2480–2492. [CrossRef] [PubMed]
16. Carlson, B.A.; Lee, B.J.; Tsuji, P.A.; Tobe, R.; Park, J.M.; Schweizer, U.; Gladyshev, V.N.; Hatfield, D.L. Selenocysteine tRNA [Ser]Sec: From Nonsense Suppressor tRNA to the Quintessential Constituent in Selenoprotein Biosynthesis. In *Selenium: Its Molecular Biology and Role in Human Health*, 4th ed.; Hatfield, D.L., Tsuji, P.A., Gladyshev, V.N., Eds.; Springer Science+Business Media, LLC: New York, NY, USA, 2016; pp. 3–12.

17. Bulteau, A.-L.; Chavatte, L. Update on selenoprotein biosynthesis. *Antioxid. Redox Signal.* **2015**, *23*, 775–794. [CrossRef] [PubMed]

18. Donovan, J.; Copeland, P.R. Threading the needle: Getting selenocysteine into proteins. *Antioxid. Redox Signal.* **2010**, *12*, 881–892. [CrossRef] [PubMed]

19. Squires, J.E.; Berry, M.J. Eukaryotic selenoprotein synthesis: Mechanistic insight incorporating new factors and new functions for old factors. *IUBMB Life* **2008**, *60*, 232–235. [CrossRef] [PubMed]

20. Papp, L.V.; Lu, J.; Holmgren, A.; Khanna, K.K. From selenium to selenoproteins: Synthesis, identity, and their role in human health. *Antioxid. Redox Signal.* **2007**, *9*, 775–806. [CrossRef] [PubMed]

21. Allmang, C.; Krol, A. Selenoprotein synthesis: UGA does not end the story. *Biochimie* **2006**, *88*, 1561–1571. [CrossRef] [PubMed]

22. Driscoll, D.M.; Copeland, P.R. Mechanism and regulation of selenoprotein synthesis. *Annu. Rev. Nutr.* **2003**, *23*, 17–40. [CrossRef] [PubMed]

23. Hatfield, D.L.; Gladyshev, V.N. How selenium has altered our understanding of the genetic code. *Mol. Cell. Biol.* **2002**, *22*, 3565–3576. [CrossRef] [PubMed]

24. Labunskyy, V.M.; Hatfield, D.L.; Gladyshev, V.N. Selenoproteins: Molecular pathways and physiological roles. *Physiol. Rev.* **2014**, *94*, 739–777. [CrossRef]

25. Brigelius-Flohe, R.; Maiorino, M. Glutathione peroxidases. *Biochim. Biophys. Acta* **2012**, *1830*, 3289–3303. [CrossRef] [PubMed]

26. Dagnell, M.; Schmidt, E.E.; Arner, E.S.J. The A to Z of modulated cell patterning by mammalian thioredoxin reductases. *Free Radic. Biol. Med.* **2018**, *115*, 484–496. [CrossRef] [PubMed]

27. Arner, E.S.J. Selective Evaluation of Thioredoxin Reductase Enzymatic Activities. *Methods Mol. Biol.* **2018**, *1661*, 301–309. [CrossRef] [PubMed]

28. Arner, E.S. Focus on mammalian thioredoxin reductases—Important selenoproteins with versatile functions. *Biochim. Biophys. Acta* **2009**, *1790*, 495–526. [CrossRef] [PubMed]

29. Arner, E.S.; Holmgren, A. Physiological functions of thioredoxin and thioredoxin reductase. *Eur. J. Biochem.* **2000**, *267*, 6102–6109. [CrossRef] [PubMed]

30. Fomenko, D.E.; Novoselov, S.V.; Natarajan, S.K.; Lee, B.C.; Koc, A.; Carlson, B.A.; Lee, T.H.; Kim, H.Y.; Hatfield, D.L.; Gladyshev, V.N. MsrB1 (methionine-R-sulfoxide reductase 1) knock-out mice: Roles of MsrB1 in redox regulation and identification of a novel selenoprotein form. *J. Biol. Chem.* **2009**, *284*, 5986–5993. [CrossRef] [PubMed]

31. Rocca, C.; Pasqua, T.; Boukhzar, L.; Anouar, Y.; Angelone, T. Progress in the emerging role of selenoproteins in cardiovascular disease: Focus on endoplasmic reticulum-resident selenoproteins. *Cell Mol. Life Sci.* **2019**. [CrossRef] [PubMed]

32. Pitts, M.W.; Hoffmann, P.R. Endoplasmic reticulum-resident selenoproteins as regulators of calcium signaling and homeostasis. *Cell Calcium* **2018**, *70*, 76–86. [CrossRef]

33. Addinsall, A.B.; Wright, C.R.; Andrikopoulos, S.; van der Poel, C.; Stupka, N. Emerging roles of endoplasmic reticulum-resident selenoproteins in the regulation of cellular stress responses and the implications for metabolic disease. *Biochem. J.* **2018**, *475*, 1037–1057. [CrossRef]

34. Molteni, C.G.; Principi, N.; Esposito, S. Reactive oxygen and nitrogen species during viral infections. *Free Radic Res.* **2014**, *48*, 1163–1169. [CrossRef]

35. Ray, P.D.; Huang, B.W.; Tsuji, Y. Reactive oxygen species (ROS) homeostasis and redox regulation in cellular signaling. *Cell. Signal.* **2012**, *24*, 981–990. [CrossRef]

36. Trachootham, D.; Alexandre, J.; Huang, P. Targeting cancer cells by ROS-mediated mechanisms: A radical therapeutic approach? *Nat. Rev. Drug Discov.* **2009**, *8*, 579–591. [CrossRef] [PubMed]

37. Andersen, J.K. Oxidative stress in neurodegeneration: Cause or consequence? *Nat. Med.* **2004**, *10* (Suppl.), S18–S25. [CrossRef] [PubMed]

38. Shukla, V.; Mishra, S.K.; Pant, H.C. Oxidative stress in neurodegeneration. *Adv. Pharmacol. Sci.* **2011**, *2011*, 572634. [CrossRef] [PubMed]

39. Paravicini, T.M.; Touyz, R.M. Redox signaling in hypertension. *Cardiovasc. Res.* **2006**, *71*, 247–258. [CrossRef] [PubMed]

40. Haigis, M.C.; Yankner, B.A. The aging stress response. *Mol. Cell* **2010**, *40*, 333–344. [CrossRef] [PubMed]

41. Khomich, O.A.; Kochetkov, S.N.; Bartosch, B.; Ivanov, A.V. Redox Biology of Respiratory Viral Infections. *Viruses* **2018**, *10*, 392. [CrossRef] [PubMed]

42. Rada, B.; Leto, T.L. Oxidative innate immune defenses by Nox/Duox family NADPH oxidases. *Contrib. Microbiol.* **2008**, *15*, 164–187. [CrossRef] [PubMed]

43. Hurst, J.K. What really happens in the neutrophil phagosome? *Free Radic Biol. Med.* **2012**, *53*, 508–520. [CrossRef] [PubMed]

44. Segal, A.W. The function of the NADPH oxidase of phagocytes and its relationship to other NOXs in plants, invertebrates, and mammals. *Int. J. Biochem. Cell Biol.* **2008**, *40*, 604–618. [CrossRef] [PubMed]

45. Jackson, S.H.; Devadas, S.; Kwon, J.; Pinto, L.A.; Williams, M.S. T cells express a phagocyte-type NADPH oxidase that is activated after T cell receptor stimulation. *Nat. Immunol.* **2004**, *5*, 818–827. [CrossRef] [PubMed]

46. Kwon, J.; Shatynski, K.E.; Chen, H.; Morand, S.; de Deken, X.; Miot, F.; Leto, T.L.; Williams, M.S. The nonphagocytic NADPH oxidase Duox1 mediates a positive feedback loop during T cell receptor signaling. *Sci. Signal.* **2010**, *3*, ra59. [CrossRef] [PubMed]

47. Thayer, T.C.; Delano, M.; Liu, C.; Chen, J.; Padgett, L.E.; Tse, H.M.; Annamali, M.; Piganelli, J.D.; Moldawer, L.L.; Mathews, C.E. Superoxide production by macrophages and T cells is critical for the induction of autoreactivity and type 1 diabetes. *Diabetes* **2011**, *60*, 2144–2151. [CrossRef] [PubMed]

48. Tse, H.M.; Thayer, T.C.; Steele, C.; Cuda, C.M.; Morel, L.; Piganelli, J.D.; Mathews, C.E. NADPH oxidase deficiency regulates Th lineage commitment and modulates autoimmunity. *J. Immunol.* **2010**, *185*, 5247–5258. [CrossRef] [PubMed]

49. Sena, L.A.; Li, S.; Jairaman, A.; Prakriya, M.; Ezponda, T.; Hildeman, D.A.; Wang, C.R.; Schumacker, P.T.; Licht, J.D.; Perlman, H.; et al. Mitochondria are required for antigen-specific T cell activation through reactive oxygen species signaling. *Immunity* **2013**, *38*, 225–236. [CrossRef] [PubMed]

50. Ivanov, A.V.; Valuev-Elliston, V.T.; Ivanova, O.N.; Kochetkov, S.N.; Starodubova, E.S.; Bartosch, B.; Isaguliants, M.G. Oxidative Stress during HIV Infection: Mechanisms and Consequences. *Oxid. Med. Cell. Longev.* **2016**, *2016*, 8910396. [CrossRef] [PubMed]

51. Reshi, M.L.; Su, Y.-C.; Hong, J.-R. RNA Viruses: ROS-Mediated Cell Death. *Int. J. Cell Biol.* **2014**, *2014*. [CrossRef] [PubMed]

52. Peterhans, E. Reactive oxygen species and nitric oxide in viral diseases. *Biol. Trace Elem. Res.* **1997**, *56*, 107–116. [CrossRef] [PubMed]

53. Beck, M.A.; Levander, O.A.; Handy, J. Selenium deficiency and viral infection. *J. Nutr.* **2003**, *133*, 1463S–1467S. [CrossRef] [PubMed]

54. Baruchel, S.; Wainberg, M.A. The role of oxidative stress in disease progression in individuals infected by the human immunodeficiency virus. *J. Leukoc. Biol.* **1992**, *52*, 111–114. [CrossRef]

55. Casola, A.; Burger, N.; Liu, T.; Jamaluddin, M.; Brasier, A.R.; Garofalo, R.P. Oxidant tone regulates RANTES gene expression in airway epithelial cells infected with respiratory syncytial virus. Role in viral-induced interferon regulatory factor activation. *J. Biol. Chem.* **2001**, *276*, 19715–19722. [CrossRef] [PubMed]

56. Jamaluddin, M.; Tian, B.; Boldogh, I.; Garofalo, R.P.; Brasier, A.R. Respiratory syncytial virus infection induces a reactive oxygen species-MSK1-phospho-Ser-276 RelA pathway required for cytokine expression. *J. Virol.* **2009**, *83*, 10605–10615. [CrossRef] [PubMed]

57. Korenaga, M.; Wang, T.; Li, Y.; Showalter, L.A.; Chan, T.; Sun, J.; Weinman, S.A. Hepatitis C virus core protein inhibits mitochondrial electron transport and increases reactive oxygen species (ROS) production. *J. Biol. Chem.* **2005**, *280*, 37481–37488. [CrossRef] [PubMed]

58. Seet, R.C.; Lee, C.Y.; Lim, E.C.; Quek, A.M.; Yeo, L.L.; Huang, S.H.; Halliwell, B. Oxidative damage in dengue fever. *Free Radic. Biol. Med.* **2009**, *47*, 375–380. [CrossRef] [PubMed]

59. Wang, Y.; Oberley, L.W.; Murhammer, D.W. Evidence of oxidative stress following the viral infection of two lepidopteran insect cell lines. *Free Radic. Biol. Med.* **2001**, *31*, 1448–1455. [CrossRef]

60. Waris, G.; Siddiqui, A. Hepatitis C virus stimulates the expression of cyclooxygenase-2 via oxidative stress: Role of prostaglandin E2 in RNA replication. *J. Virol.* **2005**, *79*, 9725–9734. [CrossRef] [PubMed]

61. Ciriolo, M.R.; Palamara, A.T.; Incerpi, S.; Lafavia, E.; Bue, M.C.; De Vito, P.; Garaci, E.; Rotilio, G. Loss of GSH, oxidative stress, and decrease of intracellular pH as sequential steps in viral infection. *J. Biol. Chem.* **1997**, *272*, 2700–2708. [CrossRef] [PubMed]

62. Garaci, E.; Palamara, A.T.; Ciriolo, M.R.; D'Agostini, C.; Abdel-Latif, M.S.; Aquaro, S.; Lafavia, E.; Rotilio, G. Intracellular GSH content and HIV replication in human macrophages. *J. Leukoc. Biol.* **1997**, *62*, 54–59. [CrossRef] [PubMed]

63. Palamara, A.T.; Perno, C.F.; Ciriolo, M.R.; Dini, L.; Balestra, E.; D'Agostini, C.; Di Francesco, P.; Favalli, C.; Rotilio, G.; Garaci, E. Evidence for antiviral activity of glutathione: In vitro inhibition of herpes simplex virus type 1 replication. *Antivir. Res.* **1995**, *27*, 237–253. [CrossRef]

64. Flores, S.C.; Marecki, J.C.; Harper, K.P.; Bose, S.K.; Nelson, S.K.; McCord, J.M. Tat protein of human immunodeficiency virus type 1 represses expression of manganese superoxide dismutase in HeLa cells. *Proc. Natl. Acad. Sci. USA* **1993**, *90*, 7632–7636. [CrossRef] [PubMed]

65. Staal, F.J.; Roederer, M.; Herzenberg, L.A.; Herzenberg, L.A. Intracellular thiols regulate activation of nuclear factor kappa B and transcription of human immunodeficiency virus. *Proc. Natl. Acad. Sci. USA* **1990**, *87*, 9943–9947. [CrossRef] [PubMed]

66. Li, C. Selenium deficiency and endemic heart failure in China: A case study of biogeochemistry for human health. *Ambio* **2007**, *36*, 90–93. [CrossRef]

67. Touat-Hamici, Z.; Bulteau, A.L.; Bianga, J.; Jean-Jacques, H.; Szpunar, J.; Lobinski, R.; Chavatte, L. Selenium-regulated hierarchy of human selenoproteome in cancerous and immortalized cells lines. *Biochim. Biophys. Acta Gen. Subj.* **2018**. [CrossRef] [PubMed]

68. Touat-Hamici, Z.; Legrain, Y.; Bulteau, A.-L.; Chavatte, L. Selective up-regulation of human selenoproteins in response to oxidative stress. *J. Biol. Chem.* **2014**, *289*, 14750–14761. [CrossRef] [PubMed]

69. Legrain, Y.; Touat-Hamici, Z.; Chavatte, L. Interplay between selenium levels, selenoprotein expression, and replicative senescence in WI-38 human fibroblasts. *J. Biol. Chem.* **2014**, *289*, 6299–6310. [CrossRef] [PubMed]

70. Latreche, L.; Duhieu, S.; Touat-Hamici, Z.; Jean-Jean, O.; Chavatte, L. The differential expression of glutathione peroxidase 1 and 4 depends on the nature of the SECIS element. *RNA Biol.* **2012**, *9*, 681–690. [CrossRef]

71. Kuhbacher, M.; Bartel, J.; Hoppe, B.; Alber, D.; Bukalis, G.; Brauer, A.U.; Behne, D.; Kyriakopoulos, A. The brain selenoproteome: Priorities in the hierarchy and different levels of selenium homeostasis in the brain of selenium-deficient rats. *J. Neurochem.* **2009**, *110*, 133–142. [CrossRef] [PubMed]

72. Latreche, L.; Jean-Jean, O.; Driscoll, D.M.; Chavatte, L. Novel structural determinants in human SECIS elements modulate the translational recoding of UGA as selenocysteine. *Nucleic Acids Res.* **2009**, *37*, 5868–5880. [CrossRef]

73. Lin, H.C.; Ho, S.C.; Chen, Y.Y.; Khoo, K.H.; Hsu, P.H.; Yen, H.C. SELENOPROTEINS. CRL2 aids elimination of truncated selenoproteins produced by failed UGA/Sec decoding. *Science* **2015**, *349*, 91–95. [CrossRef]

74. Howard, M.T.; Carlson, B.A.; Anderson, C.B.; Hatfield, D.L. Translational redefinition of UGA codons is regulated by selenium availability. *J. Biol. Chem.* **2013**, *288*, 19401–19413. [CrossRef]

75. Dalley, B.K.; Baird, L.; Howard, M.T. Studying Selenoprotein mRNA Translation Using RNA-Seq and Ribosome Profiling. *Methods Mol. Biol.* **2018**, *1661*, 103–123. [CrossRef]

76. Zhao, W.; Bohleber, S.; Schmidt, H.; Seeher, S.; Howard, M.T.; Braun, D.; Arndt, S.; Reuter, U.; Wende, H.; Birchmeier, C.; et al. Ribosome profiling of selenoproteins in vivo reveals consequences of pathogenic Secisbp2 missense mutations. *J. Biol. Chem.* **2019**. [CrossRef] [PubMed]

77. Mehta, A.; Rebsch, C.M.; Kinzy, S.A.; Fletcher, J.E.; Copeland, P.R. Efficiency of mammalian selenocysteine incorporation. *J. Biol. Chem.* **2004**, *279*, 37852–37859. [CrossRef] [PubMed]

78. Low, S.C.; Grundner-Culemann, E.; Harney, J.W.; Berry, M.J. SECIS-SBP2 interactions dictate selenocysteine incorporation efficiency and selenoprotein hierarchy. *EMBO J.* **2000**, *19*, 6882–6890. [CrossRef] [PubMed]

79. Mariotti, M.; Ridge, P.G.; Zhang, Y.; Lobanov, A.V.; Pringle, T.H.; Guigo, R.; Hatfield, D.L.; Gladyshev, V.N. Composition and evolution of the vertebrate and mammalian selenoproteomes. *PLoS ONE* **2012**, *7*, e33066. [CrossRef] [PubMed]

80. Lobanov, A.V.; Hatfield, D.L.; Gladyshev, V.N. Eukaryotic selenoproteins and selenoproteomes. *Biochim. Biophys. Acta* **2009**, *1790*, 1424–1428. [CrossRef] [PubMed]

81. Kryukov, G.V.; Castellano, S.; Novoselov, S.V.; Lobanov, A.V.; Zehtab, O.; Guigo, R.; Gladyshev, V.N. Characterization of mammalian selenoproteomes. *Science* **2003**, *300*, 1439–1443. [CrossRef] [PubMed]

82. Takebe, G.; Yarimizu, J.; Saito, Y.; Hayashi, T.; Nakamura, H.; Yodoi, J.; Nagasawa, S.; Takahashi, K. A comparative study on the hydroperoxide and thiol specificity of the glutathione peroxidase family and selenoprotein P. *J. Biol. Chem.* **2002**, *277*, 41254–41258. [CrossRef]

83. Toppo, S.; Flohe, L.; Ursini, F.; Vanin, S.; Maiorino, M. Catalytic mechanisms and specificities of glutathione peroxidases: Variations of a basic scheme. *Biochim. Biophys. Acta* **2009**, *1790*, 1486–1500. [CrossRef] [PubMed]

84. Yant, L.J.; Ran, Q.; Rao, L.; Van Remmen, H.; Shibatani, T.; Belter, J.G.; Motta, L.; Richardson, A.; Prolla, T.A. The selenoprotein GPX4 is essential for mouse development and protects from radiation and oxidative damage insults. *Free Radic. Biol. Med.* **2003**, *34*, 496–502. [CrossRef]

85. Cheng, W.H.; Ho, Y.S.; Ross, D.A.; Valentine, B.A.; Combs, G.F.; Lei, X.G. Cellular glutathione peroxidase knockout mice express normal levels of selenium-dependent plasma and phospholipid hydroperoxide glutathione peroxidases in various tissues. *J. Nutr.* **1997**, *127*, 1445–1450. [CrossRef]

86. Esworthy, R.S.; Aranda, R.; Martin, M.G.; Doroshow, J.H.; Binder, S.W.; Chu, F.F. Mice with combined disruption of Gpx1 and Gpx2 genes have colitis. *Am. J. Physiol. Gastrointest. Liver Physiol.* **2001**, *281*, G848–G855. [CrossRef]

87. Ho, Y.S.; Magnenat, J.L.; Bronson, R.T.; Cao, J.; Gargano, M.; Sugawara, M.; Funk, C.D. Mice deficient in cellular glutathione peroxidase develop normally and show no increased sensitivity to hyperoxia. *J. Biol. Chem.* **1997**, *272*, 16644–16651. [CrossRef] [PubMed]

88. De Haan, J.B.; Bladier, C.; Griffiths, P.; Kelner, M.; O'Shea, R.D.; Cheung, N.S.; Bronson, R.T.; Silvestro, M.J.; Wild, S.; Zheng, S.S.; et al. Mice with a homozygous null mutation for the most abundant glutathione peroxidase, Gpx1, show increased susceptibility to the oxidative stress-inducing agents paraquat and hydrogen peroxide. *J. Biol. Chem.* **1998**, *273*, 22528–22536. [CrossRef] [PubMed]

89. Cheng, W.H.; Ho, Y.S.; Valentine, B.A.; Ross, D.A.; Combs, G.F., Jr.; Lei, X.G. Cellular glutathione peroxidase is the mediator of body selenium to protect against paraquat lethality in transgenic mice. *J. Nutr.* **1998**, *128*, 1070–1076. [CrossRef]

90. Muller, M.F.; Florian, S.; Pommer, S.; Osterhoff, M.; Esworthy, R.S.; Chu, F.F.; Brigelius-Flohe, R.; Kipp, A.P. Deletion of glutathione peroxidase-2 inhibits azoxymethane-induced colon cancer development. *PLoS ONE* **2013**, *8*, e72055. [CrossRef] [PubMed]

91. Rhee, S.G. Overview on Peroxiredoxin. *Mol. Cells* **2016**, *39*, 1–5. [CrossRef]

92. Lu, J.; Holmgren, A. The thioredoxin antioxidant system. *Free Radic. Biol. Med.* **2013**. [CrossRef]

93. Beck, M.A.; Shi, Q.; Morris, V.C.; Levander, O.A. Benign coxsackievirus damages heart muscle in iron-loaded vitamin E-deficient mice. *Free Radic. Biol. Med.* **2005**, *38*, 112–116. [CrossRef]

94. Beck, M.A.; Williams-Toone, D.; Levander, O.A. Coxsackievirus B3-resistant mice become susceptible in Se/vitamin E deficiency. *Free Radic. Biol. Med.* **2003**, *34*, 1263–1270. [CrossRef]

95. Beck, M.A.; Nelson, H.K.; Shi, Q.; Van Dael, P.; Schiffrin, E.J.; Blum, S.; Barclay, D.; Levander, O.A. Selenium deficiency increases the pathology of an influenza virus infection. *FASEB J.* **2001**, *15*, 1481–1483. [CrossRef]

96. Beck, M.A.; Matthews, C.C. Micronutrients and host resistance to viral infection. *Proc. Nutr. Soc.* **2000**, *59*, 581–585. [CrossRef] [PubMed]

97. Beck, M.A. Nutritionally induced oxidative stress: Effect on viral disease. *Am. J. Clin. Nutr.* **2000**, *71*, 1676S–1681S. [CrossRef] [PubMed]

98. Beck, M.A.; Kolbeck, P.C.; Rohr, L.H.; Shi, Q.; Morris, V.C.; Levander, O.A. Benign human enterovirus becomes virulent in selenium-deficient mice. *J. Med. Virol.* **1994**, *43*, 166–170. [CrossRef] [PubMed]

99. Beck, M.A.; Shi, Q.; Morris, V.C.; Levander, O.A. Rapid genomic evolution of a non-virulent coxsackievirus B3 in selenium-deficient mice results in selection of identical virulent isolates. *Nat. Med.* **1995**, *1*, 433–436. [CrossRef] [PubMed]

100. Jubelt, B.; Lipton, H.L. Enterovirus/picornavirus infections. *Handb. Clin. Neurol.* **2014**, *123*, 379–416. [CrossRef] [PubMed]

101. Muehlenbachs, A.; Bhatnagar, J.; Zaki, S.R. Tissue tropism, pathology and pathogenesis of enterovirus infection. *J. Pathol.* **2015**, *235*, 217–228. [CrossRef] [PubMed]

102. Yang, G.Q.; Chen, J.S.; Wen, Z.M.; Ge, K.Y.; Zhu, L.Z.; Chen, X.C.; Chen, X.S. The role of selenium in Keshan disease. *Adv. Nutr. Res.* **1984**, *6*, 203–231. [PubMed]

103. Loscalzo, J. Keshan disease, selenium deficiency, and the selenoproteome. *N. Engl. J. Med.* **2014**, *370*, 1756–1760. [CrossRef] [PubMed]

104. Li, Y.; Yang, Y.; Chen, H. Detection of enteroviral RNA in paraffin-embedded myocardial tissue from patients with Keshan by nested PCR. *Zhonghua Yi Xue Za Zhi* **1995**, *75*, 344–345. [PubMed]

105. Li, Y.; Peng, T.; Yang, Y.; Niu, C.; Archard, L.C.; Zhang, H. High prevalence of enteroviral genomic sequences in myocardium from cases of endemic cardiomyopathy (Keshan disease) in China. *Heart* **2000**, *83*, 696–701. [CrossRef]

106. Liu, Y.; Chiba, M.; Inaba, Y.; Kondo, M. Keshan disease—A review from the aspect of history and etiology. *Nihon Eiseigaku Zasshi* **2002**, *56*, 641–648. [CrossRef] [PubMed]

107. Beck, M.A. Selenium and host defence towards viruses. *Proc. Nutr. Soc.* **1999**, *58*, 707–711. [CrossRef] [PubMed]

108. Beck, M.A.; Esworthy, R.S.; Ho, Y.; Chu, F. Glutathione peroxidase protects mice from viral-induced myocarditis. *FASEB J.* **1998**, *12*, 1143–1149. [CrossRef] [PubMed]

109. Beck, M.A. Rapid genomic evolution of a non-virulent coxsackievirus B3 in selenium-deficient mice. *Biomed. Environ. Sci.* **1997**, *10*, 307–315. [PubMed]

110. Beck, M.A. Increased virulence of coxsackievirus B3 in mice due to vitamin E or selenium deficiency. *J. Nutr.* **1997**, *127*, 966S–970S. [CrossRef] [PubMed]

111. Beck, M.A.; Kolbeck, P.C.; Shi, Q.; Rohr, L.H.; Morris, V.C.; Levander, O.A. Increased virulence of a human enterovirus (coxsackievirus B3) in selenium-deficient mice. *J. Infect. Dis.* **1994**, *170*, 351–357. [CrossRef] [PubMed]

112. Cooper, L.T.; Rader, V.; Ralston, N.V. The roles of selenium and mercury in the pathogenesis of viral cardiomyopathy. *Congest Heart Fail* **2007**, *13*, 193–199. [CrossRef] [PubMed]

113. Pleschka, S. Overview of influenza viruses. *Curr. Top. Microbiol. Immunol.* **2013**, *370*, 1–20. [CrossRef]

114. Lim, J.Y.; Oh, E.; Kim, Y.; Jung, W.W.; Kim, H.S.; Lee, J.; Sul, D. Enhanced oxidative damage to DNA, lipids, and proteins and levels of some antioxidant enzymes, cytokines, and heat shock proteins in patients infected with influenza H1N1 virus. *Acta Virol.* **2014**, *58*, 253–260. [CrossRef]

115. Ng, M.P.; Lee, J.C.; Loke, W.M.; Yeo, L.L.; Quek, A.M.; Lim, E.C.; Halliwell, B.; Seet, R.C. Does influenza A infection increase oxidative damage? *Antioxid. Redox Signal.* **2014**, *21*, 1025–1031. [CrossRef]

116. Erkekoglu, P.; Asci, A.; Ceyhan, M.; Kizilgun, M.; Schweizer, U.; Atas, C.; Kara, A.; Kocer Giray, B. Selenium levels, selenoenzyme activities and oxidant/antioxidant parameters in H1N1-infected children. *Turk. J. Pediatr.* **2013**, *55*, 271–282. [PubMed]

117. Buffinton, G.D.; Christen, S.; Peterhans, E.; Stocker, R. Oxidative stress in lungs of mice infected with influenza A virus. *Free Radic. Res. Commun.* **1992**, *16*, 99–110. [CrossRef] [PubMed]

118. Hennet, T.; Peterhans, E.; Stocker, R. Alterations in antioxidant defences in lung and liver of mice infected with influenza A virus. *J. Gen. Virol.* **1992**, *73 Pt 1*, 39–46. [CrossRef]

119. Amatore, D.; Sgarbanti, R.; Aquilano, K.; Baldelli, S.; Limongi, D.; Civitelli, L.; Nencioni, L.; Garaci, E.; Ciriolo, M.R.; Palamara, A.T. Influenza virus replication in lung epithelial cells depends on redox-sensitive pathways activated by NOX4-derived ROS. *Cell Microbiol.* **2015**, *17*, 131–145. [CrossRef] [PubMed]

120. Ye, S.; Lowther, S.; Stambas, J. Inhibition of reactive oxygen species production ameliorates inflammation induced by influenza A viruses via upregulation of SOCS1 and SOCS3. *J. Virol.* **2015**, *89*, 2672–2683. [CrossRef] [PubMed]

121. Nelson, H.K.; Shi, Q.; Van Dael, P.; Schiffrin, E.J.; Blum, S.; Barclay, D.; Levander, O.A.; Beck, M.A. Host nutritional selenium status as a driving force for influenza virus mutations. *FASEB J.* **2001**, *15*, 1846–1848. [CrossRef] [PubMed]

122. Jaspers, I.; Zhang, W.; Brighton, L.E.; Carson, J.L.; Styblo, M.; Beck, M.A. Selenium deficiency alters epithelial cell morphology and responses to influenza. *Free Radic. Biol. Med.* **2007**, *42*, 1826–1837. [CrossRef] [PubMed]

123. Sheridan, P.A.; Zhong, N.; Carlson, B.A.; Perella, C.M.; Hatfield, D.L.; Beck, M.A. Decreased selenoprotein expression alters the immune response during influenza virus infection in mice. *J. Nutr.* **2007**, *137*, 1466–1471. [CrossRef] [PubMed]

124. Stýblo, M.; Walton, F.S.; Harmon, A.W.; Sheridan, P.A.; Beck, M.A. Activation of superoxide dismutase in selenium-deficient mice infected with influenza virus. *J. Trace Elem. Med. Biol.* **2007**, *21*, 52–62. [CrossRef] [PubMed]

125. Li, W.; Beck, M.A. Selenium deficiency induced an altered immune response and increased survival following influenza A/Puerto Rico/8/34 infection. *Exp. Biol. Med.* **2007**, *232*, 412–419.

126. Ferguson, M.R.; Rojo, D.R.; von Lindern, J.J.; O'Brien, W.A. HIV-1 replication cycle. *Clin. Lab. Med.* **2002**, *22*, 611–635. [CrossRef]

127. Nyamweya, S.; Hegedus, A.; Jaye, A.; Rowland-Jones, S.; Flanagan, K.L.; Macallan, D.C. Comparing HIV-1 and HIV-2 infection: Lessons for viral immunopathogenesis. *Rev. Med. Virol.* **2013**, *23*, 221–240. [CrossRef] [PubMed]

128. Palella, F.J., Jr.; Delaney, K.M.; Moorman, A.C.; Loveless, M.O.; Fuhrer, J.; Satten, G.A.; Aschman, D.J.; Holmberg, S.D. Declining morbidity and mortality among patients with advanced human immunodeficiency virus infection. HIV Outpatient Study Investigators. *N. Engl. J. Med.* **1998**, *338*, 853–860. [CrossRef] [PubMed]

129. Samji, H.; Cescon, A.; Hogg, R.S.; Modur, S.P.; Althoff, K.N.; Buchacz, K.; Burchell, A.N.; Cohen, M.; Gebo, K.A.; Gill, M.J.; et al. Closing the gap: Increases in life expectancy among treated HIV-positive individuals in the United States and Canada. *PLoS ONE* **2013**, *8*, e81355. [CrossRef] [PubMed]

130. Pace, G.W.; Leaf, C.D. The role of oxidative stress in HIV disease. *Free Radic. Biol. Med.* **1995**, *19*, 523–528. [CrossRef]

131. Hoffmann, F.W.; Hashimoto, A.C.; Shafer, L.A.; Dow, S.; Berry, M.J.; Hoffmann, P.R. Dietary selenium modulates activation and differentiation of CD4+ T cells in mice through a mechanism involving cellular free thiols. *J. Nutr.* **2010**, *140*, 1155–1161. [CrossRef]

132. Huang, Z.; Rose, A.H.; Hoffmann, P.R. The role of selenium in inflammation and immunity: From molecular mechanisms to therapeutic opportunities. *Antioxid. Redox Signal.* **2012**, *16*, 705–743. [CrossRef]

133. Bogden, J.D.; Oleske, J.M. The essential trace minerals, immunity, and progression of HIV-1 infection. *Nutr. Res.* **2007**, *27*, 69–77. [CrossRef]

134. Stone, C.A.; Kawai, K.; Kupka, R.; Fawzi, W.W. Role of selenium in HIV infection. *Nutr. Rev.* **2010**, *68*, 671–681. [CrossRef]

135. Chandrasekhar, A.; Gupta, A. Nutrition and disease progression pre-highly active antiretroviral therapy (HAART) and post-HAART: Can good nutrition delay time to HAART and affect response to HAART? *Am. J. Clin. Nutr.* **2011**, *94*, 1703S–1715S. [CrossRef]

136. Pitney, C.L.; Royal, M.; Klebert, M. Selenium supplementation in HIV-infected patients: Is there any potential clinical benefit? *J. Assoc. Nurses AIDS Care* **2009**, *20*, 326–333. [CrossRef] [PubMed]

137. Hurwitz, B.E.; Klaus, J.R.; Llabre, M.M.; Gonzalez, A.; Lawrence, P.J.; Maher, K.J.; Greeson, J.M.; Baum, M.K.; Shor-Posner, G.; Skyler, J.S.; et al. Suppression of human immunodeficiency virus type 1 viral load with selenium supplementation: A randomized controlled trial. *Arch. Intern. Med.* **2007**, *167*, 148–154. [CrossRef] [PubMed]

138. Baum, M.K.; Campa, A.; Lai, S.; Sales Martinez, S.; Tsalaile, L.; Burns, P.; Farahani, M.; Li, Y.; van Widenfelt, E.; Page, J.B.; et al. Effect of micronutrient supplementation on disease progression in asymptomatic, antiretroviral-naive, HIV-infected adults in Botswana: A randomized clinical trial. *JAMA* **2013**, *310*, 2154–2163. [CrossRef] [PubMed]

139. Kamwesiga, J.; Mutabazi, V.; Kayumba, J.; Tayari, J.C.; Uwimbabazi, J.C.; Batanage, G.; Uwera, G.; Baziruwiha, M.; Ntizimira, C.; Murebwayire, A.; et al. Effect of selenium supplementation on CD4+ T-cell recovery, viral suppression and morbidity of HIV-infected patients in Rwanda: A randomized controlled trial. *AIDS* **2015**, *29*, 1045–1052. [CrossRef] [PubMed]

140. Williams, A.A.; Sitole, L.J.; Meyer, D. HIV/HAART-associated oxidative stress is detectable by metabonomics. *Mol. Biosyst.* **2017**, *13*, 2202–2217. [CrossRef] [PubMed]

141. De Menezes Barbosa, E.G.; Junior, F.B.; Machado, A.A.; Navarro, A.M. A longer time of exposure to antiretroviral therapy improves selenium levels. *Clin. Nutr.* **2015**, *34*, 248–251. [CrossRef]

142. Gladyshev, V.N.; Stadtman, T.C.; Hatfield, D.L.; Jeang, K.T. Levels of major selenoproteins in T cells decrease during HIV infection and low molecular mass selenium compounds increase. *Proc. Natl. Acad. Sci. USA* **1999**, *96*, 835–839. [CrossRef]

143. Davis, G.L. Treatment of chronic hepatitis C. *BMJ* **2001**, *323*, 1141–1142. [CrossRef]

144. Ivanov, A.V.; Bartosch, B.; Smirnova, O.A.; Isaguliants, M.G.; Kochetkov, S.N. HCV and oxidative stress in the liver. *Viruses* **2013**, *5*, 439–469. [CrossRef]

145. Boya, P.; de la Pena, A.; Beloqui, O.; Larrea, E.; Conchillo, M.; Castelruiz, Y.; Civeira, M.P.; Prieto, J. Antioxidant status and glutathione metabolism in peripheral blood mononuclear cells from patients with chronic hepatitis C. *J. Hepatol.* **1999**, *31*, 808–814. [CrossRef]

146. De Maria, N.; Colantoni, A.; Fagiuoli, S.; Liu, G.J.; Rogers, B.K.; Farinati, F.; Van Thiel, D.H.; Floyd, R.A. Association between reactive oxygen species and disease activity in chronic hepatitis C. *Free Radic. Biol. Med.* **1996**, *21*, 291–295. [CrossRef]

147. Bianchi, G.; Marchesini, G.; Brizi, M.; Rossi, B.; Forlani, G.; Boni, P.; Melchionda, N.; Thomaseth, K.; Pacini, G. Nutritional effects of oral zinc supplementation in cirrhosis. *Nutr. Res.* **2000**, *20*, 1079–1089. [CrossRef]

148. Ivanov, A.V.; Smirnova, O.A.; Ivanova, O.N.; Masalova, O.V.; Kochetkov, S.N.; Isaguliants, M.G. Hepatitis C virus proteins activate NRF2/ARE pathway by distinct ROS-dependent and independent mechanisms in HUH7 cells. *PLoS ONE* **2011**, *6*, e24957. [CrossRef] [PubMed]

149. Pal, S.; Polyak, S.J.; Bano, N.; Qiu, W.C.; Carithers, R.L.; Shuhart, M.; Gretch, D.R.; Das, A. Hepatitis C virus induces oxidative stress, DNA damage and modulates the DNA repair enzyme NEIL1. *J. Gastroenterol. Hepatol.* **2010**, *25*, 627–634. [CrossRef] [PubMed]

150. Bureau, C.; Bernad, J.; Chaouche, N.; Orfila, C.; Beraud, M.; Gonindard, C.; Alric, L.; Vinel, J.P.; Pipy, B. Nonstructural 3 protein of hepatitis C virus triggers an oxidative burst in human monocytes via activation of NADPH oxidase. *J. Biol. Chem.* **2001**, *276*, 23077–23083. [CrossRef] [PubMed]

151. Garcia-Mediavilla, M.V.; Sanchez-Campos, S.; Gonzalez-Perez, P.; Gomez-Gonzalo, M.; Majano, P.L.; Lopez-Cabrera, M.; Clemente, G.; Garcia-Monzon, C.; Gonzalez-Gallego, J. Differential contribution of hepatitis C virus NS5A and core proteins to the induction of oxidative and nitrosative stress in human hepatocyte-derived cells. *J. Hepatol.* **2005**, *43*, 606–613. [CrossRef] [PubMed]

152. Ko, W.S.; Guo, C.H.; Yeh, M.S.; Lin, L.Y.; Hsu, G.S.; Chen, P.C.; Luo, M.C.; Lin, C.Y. Blood micronutrient, oxidative stress, and viral load in patients with chronic hepatitis C. *World J. Gastroenterol.* **2005**, *11*, 4697–4702. [CrossRef] [PubMed]

153. Look, M.P.; Rockstroh, J.K.; Rao, G.S.; Kreuzer, K.A.; Barton, S.; Lemoch, H.; Sudhop, T.; Hoch, J.; Stockinger, K.; Spengler, U.; et al. Serum selenium, plasma glutathione (GSH) and erythrocyte glutathione peroxidase (GSH-Px)-levels in asymptomatic versus symptomatic human immunodeficiency virus-1 (HIV-1)-infection. *Eur. J. Clin. Nutr.* **1997**, *51*, 266–272. [CrossRef]

154. Chusri, P.; Kumthip, K.; Hong, J.; Zhu, C.; Duan, X.; Jilg, N.; Fusco, D.N.; Brisac, C.; Schaefer, E.A.; Cai, D.; et al. HCV induces transforming growth factor beta1 through activation of endoplasmic reticulum stress and the unfolded protein response. *Sci. Rep.* **2016**, *6*, 22487. [CrossRef]

155. Guerriero, E.; Accardo, M.; Capone, F.; Colonna, G.; Castello, G.; Costantini, S. Assessment of the Selenoprotein M (SELM) over-expression on human hepatocellular carcinoma tissues by immunohistochemistry. *Eur. J. Histochem.* **2014**, *58*, 2433. [CrossRef]

156. Tsai, K.N.; Kuo, C.F.; Ou, J.J. Mechanisms of Hepatitis B Virus Persistence. *Trends Microbiol.* **2018**, *26*, 33–42. [CrossRef] [PubMed]

157. Karayiannis, P. Hepatitis B virus: Virology, molecular biology, life cycle and intrahepatic spread. *Hepatol. Int.* **2017**, *11*, 500–508. [CrossRef] [PubMed]

158. Balamtekin, N.; Kurekci, A.E.; Atay, A.; Kalman, S.; Okutan, V.; Gokcay, E.; Aydin, A.; Sener, K.; Safali, M.; Ozcan, O. Plasma levels of trace elements have an implication on interferon treatment of children with chronic hepatitis B infection. *Biol. Trace Elem. Res.* **2010**, *135*, 153–161. [CrossRef] [PubMed]

159. Abediankenari, S.; Ghasemi, M.; Nasehi, M.M.; Abedi, S.; Hosseini, V. Determination of trace elements in patients with chronic hepatitis B. *Acta Med. Iran.* **2011**, *49*, 667–669. [PubMed]

160. Khan, M.S.; Dilawar, S.; Ali, I.; Rauf, N. The possible role of selenium concentration in hepatitis B and C patients. *Saudi J. Gastroenterol. Off. J. Saudi Gastroenterol. Assoc.* **2012**, *18*, 106–110. [CrossRef] [PubMed]

161. Cheng, Z.; Zhi, X.; Sun, G.; Guo, W.; Huang, Y.; Sun, W.; Tian, X.; Zhao, F.; Hu, K. Sodium selenite suppresses hepatitis B virus transcription and replication in human hepatoma cell lines. *J. Med. Virol.* **2016**, *88*, 653–663. [CrossRef]

162. Yu, S.Y.; Zhu, Y.J.; Li, W.G. Protective role of selenium against hepatitis B virus and primary liver cancer in Qidong. *Biol. Trace Elem. Res.* **1997**, *56*, 117–124. [CrossRef]

163. Finsterbusch, T.; Mankertz, A. Porcine circoviruses—Small but powerful. *Virus Res.* **2009**, *143*, 177–183. [CrossRef]

164. Pan, Q.; Huang, K.; He, K.; Lu, F. Effect of different selenium sources and levels on porcine circovirus type 2 replication in vitro. *J. Trace Elem. Med. Biol. Organ Soc. Miner. Trace Elem. (GMS)* **2008**, *22*, 143–148. [CrossRef]

165. Chen, X.; Ren, F.; Hesketh, J.; Shi, X.; Li, J.; Gan, F.; Huang, K. Selenium blocks porcine circovirus type 2 replication promotion induced by oxidative stress by improving GPx1 expression. *Free Radic. Biol. Med.* **2012**, *53*, 395–405. [CrossRef]

166. Gan, F.; Hu, Z.; Huang, Y.; Xue, H.; Huang, D.; Qian, G.; Hu, J.; Chen, X.; Wang, T.; Huang, K. Overexpression of pig selenoprotein S blocks OTA-induced promotion of PCV2 replication by inhibiting oxidative stress and p38 phosphorylation in PK15 cells. *Oncotarget* **2016**, *7*, 20469–20485. [CrossRef] [PubMed]

167. Liu, D.; Xu, J.; Qian, G.; Hamid, M.; Gan, F.; Chen, X.; Huang, K. Selenizing astragalus polysaccharide attenuates PCV2 replication promotion caused by oxidative stress through autophagy inhibition via PI3K/AKT activation. *Int. J. Biol. Macromol.* **2018**, *108*, 350–359. [CrossRef] [PubMed]

168. Qian, G.; Liu, D.; Hu, J.; Gan, F.; Hou, L.; Zhai, N.; Chen, X.; Huang, K. SeMet attenuates OTA-induced PCV2 replication promotion by inhibiting autophagy by activating the AKT/mTOR signaling pathway. *Vet. Res.* **2018**, *49*, 15. [CrossRef]

169. Sumba, P.O.; Kabiru, E.W.; Namuyenga, E.; Fiore, N.; Otieno, R.O.; Moormann, A.M.; Orago, A.S.; Rosenbaum, P.F.; Rochford, R. Microgeographic variations in Burkitt's lymphoma incidence correlate with differences in malnutrition, malaria and Epstein-Barr virus. *Br. J. Cancer* **2010**, *103*, 1736–1741. [CrossRef]

170. Jian, S.-W.; Mei, C.-E.; Liang, Y.-N.; Li, D.; Chen, Q.-L.; Luo, H.-L.; Li, Y.-Q.; Cai, T.-Y. Influence of selenium-rich rice on transformation of umbilical blood B lymphocytes by Epstein-Barr virus and Epstein-Barr virus early antigen expression. *Ai Zheng = Aizheng = Chin. J. Cancer* **2003**, *22*, 26–29.

171. Taylor, E.W.; Nadimpalli, R.G.; Ramanathan, C.S. Genomic structures of viral agents in relation to the biosynthesis of selenoproteins. *Biol. Elem. Res.* **1997**, *56*, 63–91. [CrossRef] [PubMed]

172. De Luca, C.; Kharaeva, Z.; Raskovic, D.; Pastore, P.; Luci, A.; Korkina, L. Coenzyme Q(10), vitamin E, selenium, and methionine in the treatment of chronic recurrent viral mucocutaneous infections. *Nutrition (Burbank)* **2012**, *28*, 509–514. [CrossRef]

173. Sartori, G.; Jardim, N.S.; Marcondes Sari, M.H.; Dobrachinski, F.; Pesarico, A.P.; Rodrigues, L.C.; Cargnelutti, J.; Flores, E.F.; Prigol, M.; Nogueira, C.W. Antiviral Action of Diphenyl Diselenide on Herpes Simplex Virus 2 Infection in Female BALB/c Mice. *J. Cell. Biochem.* **2016**, *117*, 1638–1648. [CrossRef]

174. Reffett, J.K.; Spears, J.W.; Brown, T.T. Effect of dietary selenium on the primary and secondary immune response in calves challenged with infectious bovine rhinotracheitis virus. *J. Nutr.* **1988**, *118*, 229–235. [CrossRef]

175. Shisler, J.L.; Senkevich, T.G.; Berry, M.J.; Moss, B. Ultraviolet-induced cell death blocked by a selenoprotein from a human dermatotropic poxvirus. *Science* **1998**, *279*, 102–105. [CrossRef]

176. Mix, H.; Lobanov, A.V.; Gladyshev, V.N. SECIS elements in the coding regions of selenoprotein transcripts are functional in higher eukaryotes. *Nucleic Acids Res.* **2007**, *35*, 414–423. [CrossRef] [PubMed]

177. Polansky, H.; Itzkovitz, E.; Javaherian, A. Human papillomavirus (HPV): Systemic treatment with Gene-Eden-VIR/Novirin safely and effectively clears virus. *Drug Des. Dev. Ther.* **2017**, *11*, 575–583. [CrossRef] [PubMed]

178. Liu, G.; Yang, G.; Guan, G.; Zhang, Y.; Ren, W.; Yin, J.; Aguilar, Y.M.; Luo, W.; Fang, J.; Yu, X.; et al. Effect of Dietary Selenium Yeast Supplementation on Porcine Circovirus Type 2 (PCV2) Infections in Mice. *PLoS ONE* **2015**, *10*. [CrossRef] [PubMed]

179. Gómez, R.M.; Berría, M.I.; Levander, O.A. Host selenium status selectively influences susceptibility to experimental viral myocarditis. *Biol. Trace Elem. Res.* **2001**, *80*, 23–31. [CrossRef]

180. Sepúlveda, R.T.; Zhang, J.; Watson, R.R. Selenium supplementation decreases coxsackievirus heart disease during murine AIDS. *Cardiovasc. Toxicol.* **2002**, *2*, 53–61. [CrossRef] [PubMed]

181. Molin, Y.; Frisk, P.; Ilbäck, N.-G. Viral RNA kinetics is associated with changes in trace elements in target organs of Coxsackie virus B3 infection. *Microbes Infect.* **2009**, *11*, 493–499. [CrossRef] [PubMed]

182. Jun, E.J.; Ye, J.S.; Hwang, I.S.; Kim, Y.K.; Lee, H. Selenium deficiency contributes to the chronic myocarditis in coxsackievirus-infected mice. *Acta Virol.* **2011**, *55*, 23–29. [CrossRef]

183. Zhang, W.; Ramanathan, C.S.; Nadimpalli, R.G.; Bhat, A.A.; Cox, A.G.; Taylor, E.W. Selenium-dependent glutathione peroxidase modules encoded by RNA viruses. *Biol. Trace Elem. Res.* **1999**, *70*, 97–116. [CrossRef]

184. Cermelli, C.; Vinceti, M.; Scaltriti, E.; Bazzani, E.; Beretti, F.; Vivoli, G.; Portolani, M. Selenite inhibition of Coxsackie virus B5 replication: Implications on the etiology of Keshan disease. *J. Trace Elem. Med. Biol. Organ Soc. Miner. Trace Elem. (GMS)* **2002**, *16*, 41–46. [CrossRef]

185. Broome, C.S.; McArdle, F.; Kyle, J.A.M.; Andrews, F.; Lowe, N.M.; Hart, C.A.; Arthur, J.R.; Jackson, M.J. An increase in selenium intake improves immune function and poliovirus handling in adults with marginal selenium status. *Am. J. Clin. Nutr.* **2004**, *80*, 154–162. [CrossRef]

186. Abou-Zeina, H.A.A.; Nasr, S.M.; Nassar, S.A.; Farag, T.K.; El-Bayoumy, M.K.; Ata, E.B.; Hassan, N.M.F.; Abdel-Aziem, S.H. Beneficial effects of antioxidants in improving health conditions of sheep infected with foot-and-mouth disease. *Trop. Anim. Health Prod.* **2019**. [CrossRef] [PubMed]

187. Sheehan, H.B.; Benetucci, J.; Muzzio, E.; Redini, L.; Naveira, J.; Segura, M.; Weissenbacher, M.; Tang, A.M. High rates of serum selenium deficiency among HIV- and HCV-infected and uninfected drug users in Buenos Aires, Argentina. *Public Health Nutr.* **2012**, *15*, 538–545. [CrossRef] [PubMed]

188. Groenbaek, K.; Friis, H.; Hansen, M.; Ring-Larsen, H.; Krarup, H.B. The effect of antioxidant supplementation on hepatitis C viral load, transaminases and oxidative status: A randomized trial among chronic hepatitis C virus-infected patients. *Eur. J. Gastroenterol. Hepatol.* **2006**, *18*, 985–989. [CrossRef] [PubMed]

189. Zhang, W.; Cox, A.G.; Taylor, E.W. Hepatitis C virus encodes a selenium-dependent glutathione peroxidase gene. Implications for oxidative stress as a risk factor in progression to hepatocellular carcinoma. *Medizinische Klinik (Munich)* **1999**, *94* (Suppl. 3), 2–6. [CrossRef] [PubMed]

190. Verma, S.; Molina, Y.; Lo, Y.Y.; Cropp, B.; Nakano, C.; Yanagihara, R.; Nerurkar, V.R. In vitro effects of selenium deficiency on West Nile virus replication and cytopathogenicity. *Virol. J.* **2008**, *5*, 66. [CrossRef] [PubMed]

191. Zhong, H.; Taylor, E.W. Structure and dynamics of a predicted ferredoxin-like selenoprotein in Japanese encephalitis virus. *J. Mol. Graph. Model.* **2004**, *23*, 223–231. [CrossRef] [PubMed]

192. Fang, L.-Q.; Goeijenbier, M.; Zuo, S.-Q.; Wang, L.-P.; Liang, S.; Klein, S.L.; Li, X.-L.; Liu, K.; Liang, L.; Gong, P.; et al. The association between hantavirus infection and selenium deficiency in mainland China. *Viruses* **2015**, *7*, 333–351. [CrossRef]

193. Al-Sonboli, N.; Al-Aghbari, N.; Al-Aryani, A.; Atef, Z.; Brabin, B.; Shenkin, A.; Roberts, E.; Harper, G.; Hart, C.A.; Cuevas, L.E. Micronutrient concentrations in respiratory syncytial virus and human metapneumovirus in Yemeni children. *Ann. Trop. Paediatr.* **2009**, *29*, 35–40. [CrossRef]

194. Liu, X.; Yin, S.; Li, G. Effects of selenium supplement on acute lower respiratory tract infection caused by respiratory syncytial virus. *Zhonghua Yu Fang Yi Xue Za Zhi [Chin. J. Prev. Med.]* **1997**, *31*, 358–361.

195. Taylor, E.W.; Ruzicka, J.A.; Premadasa, L.; Zhao, L. Cellular Selenoprotein mRNA Tethering via Antisense Interactions with Ebola and HIV-1 mRNAs May Impact Host Selenium Biochemistry. *Curr. Top Med. Chem.* **2016**, *16*, 1530–1535. [CrossRef]

196. Moya, M.; Bautista, E.G.; Velázquez-González, A.; Vázquez-Gutiérrez, F.; Tzintzun, G.; García-Arreola, M.E.; Castillejos, M.; Hernández, A. Potentially-toxic and essential elements profile of AH1N1 patients in Mexico City. *Sci. Rep.* **2013**, *3*, 1284. [CrossRef] [PubMed]

197. Yu, L.; Sun, L.; Nan, Y.; Zhu, L.-Y. Protection from H1N1 influenza virus infections in mice by supplementation with selenium: A comparison with selenium-deficient mice. *Biol. Trace Elem. Res.* **2011**, *141*, 254–261. [CrossRef] [PubMed]

198. Shojadoost, B.; Kulkarni, R.R.; Yitbarek, A.; Laursen, A.; Taha-Abdelaziz, K.; Negash Alkie, T.; Barjesteh, N.; Quinteiro-Filho, W.M.; Smith, T.K.; Sharif, S. Dietary selenium supplementation enhances antiviral immunity in chickens challenged with low pathogenic avian influenza virus subtype H9N2. *Vet. Immunol. Immunopathol.* **2019**, *207*, 62–68. [CrossRef] [PubMed]

199. Deryabin, P.G.; Lvov, D.K.; Botikov, A.G.; Ivanov, V.; Kalinovsky, T.; Niedzwiecki, A.; Rath, M. Effects of a nutrient mixture on infectious properties of the highly pathogenic strain of avian influenza virus A/H5N1. *BioFactors (Oxford)* **2008**, *33*, 85–97. [CrossRef]

200. Reffett, J.K.; Spears, J.W.; Brown, T.T. Effect of dietary selenium and vitamin E on the primary and secondary immune response in lambs challenged with parainfluenza3 virus. *J. Anim. Sci.* **1988**, *66*, 1520–1528. [CrossRef] [PubMed]

201. Favier, A.; Sappey, C.; Leclerc, P.; Faure, P.; Micoud, M. Antioxidant status and lipid peroxidation in patients infected with HIV. *Chem. Biol. Interact.* **1994**, *91*, 165–180. [CrossRef]

202. Allavena, C.; Dousset, B.; May, T.; Dubois, F.; Canton, P.; Belleville, F. Relationship of trace element, immunological markers, and HIV1 infection progression. *Biol. Trace Elem. Res.* **1995**, *47*, 133–138. [CrossRef]

203. Baum, M.K.; Shor-Posner, G.; Lai, S.; Zhang, G.; Lai, H.; Fletcher, M.A.; Sauberlich, H.; Page, J.B. High risk of HIV-related mortality is associated with selenium deficiency. *J. Acquir. Immune Defic. Syndr. Hum. Retrovirol.* **1997**, *15*, 370–374. [CrossRef]

204. Baum, M.K.; Shor-Posner, G.; Zhang, G.; Lai, H.; Quesada, J.A.; Campa, A.; Jose-Burbano, M.; Fletcher, M.A.; Sauberlich, H.; Page, J.B. HIV-1 infection in women is associated with severe nutritional deficiencies. *J. Acquir. Immune Defic. Syndr. Hum. Retrovirol.* **1997**, *16*, 272–278. [CrossRef]

205. Sabe, R.; Rubio, R.; Garcia-Beltran, L. Reference values of selenium in plasma in population from Barcelona. Comparison with several pathologies. *J. Trace Elem. Med. Biol.* **2002**, *16*, 231–237. [CrossRef]

206. Kupka, R.; Msamanga, G.I.; Spiegelman, D.; Morris, S.; Mugusi, F.; Hunter, D.J.; Fawzi, W.W. Selenium status is associated with accelerated HIV disease progression among HIV-1-infected pregnant women in Tanzania. *J. Nutr.* **2004**, *134*, 2556–2560. [CrossRef] [PubMed]

207. Ogunro, P.S.; Ogungbamigbe, T.O.; Elemie, P.O.; Egbewale, B.E.; Adewole, T.A. Plasma selenium concentration and glutathione peroxidase activity in HIV-1/AIDS infected patients: A correlation with the disease progression. *Niger Postgrad. Med. J.* **2006**, *13*, 1–5. [PubMed]

208. Khalili, H.; Soudbakhsh, A.; Hajiabdolbaghi, M.; Dashti-Khavidaki, S.; Poorzare, A.; Saeedi, A.A.; Sharififar, R. Nutritional status and serum zinc and selenium levels in Iranian HIV infected individuals. *BMC Infect. Dis.* **2008**, *8*, 165. [CrossRef] [PubMed]

209. Djinhi, J.; Tiahou, G.; Zirihi, G.; Lohoues, E.; Monde, A.; Camara, C.; Sess, E. Selenium deficiency and oxidative stress in asymptomatic HIV1-infected patients in Côte d'Ivoire. *Bull. Soc. Pathol. Exotique (1990)* **2009**, *102*, 11–13. [CrossRef]

210. Okwara, E.C.; Meludu, S.C.; Okwara, J.E.; Enwere, O.O.; Diwe, K.C.; Amah, U.K.; Ubajaka, C.F.; Chukwulebe, A.E.; Ezeugwunne, I.P. Selenium, zinc and magnesium status of HIV positive adults presenting at a university teaching hospital in Orlu-Eastern Nigeria. *Niger. J. Med. J. Natl. Assoc. Res. Doctors Niger.* **2012**, *21*, 165–168.

211. Akinboro, A.O.; Mejiuni, D.A.; Onayemi, O.; Ayodele, O.E.; Atiba, A.S.; Bamimore, G.M. Serum selenium and skin diseases among Nigerians with human immunodeficiency virus/acquired immune deficiency syndrome. *HIV/AIDS* **2013**, *5*, 215–221. [CrossRef] [PubMed]

212. Anyabolu, H.C.; Adejuyigbe, E.A.; Adeodu, O.O. Serum Micronutrient Status of Haart-Naïve, HIV Infected Children in South Western Nigeria: A Case Controlled Study. *AIDS Res. Treat.* **2014**, *2014*, 351043. [CrossRef] [PubMed]

213. Henderson, R.A.; Talusan, K.; Hutton, N.; Yolken, R.H.; Caballero, B. Serum and plasma markers of nutritional status in children infected with the human immunodeficiency virus. *J. Am. Diet. Assoc.* **1997**, *97*, 1377–1381. [CrossRef]

214. Bunupuradah, T.; Ubolyam, S.; Hansudewechakul, R.; Kosalaraksa, P.; Ngampiyaskul, C.; Kanjanavanit, S.; Wongsawat, J.; Luesomboon, W.; Pinyakorn, S.; Kerr, S.; et al. Correlation of selenium and zinc levels to antiretroviral treatment outcomes in Thai HIV-infected children without severe HIV symptoms. *Eur. J. Clin. Nutr.* **2012**, *66*, 900–905. [CrossRef] [PubMed]

215. Hileman, C.O.; Dirajlal-Fargo, S.; Lam, S.K.; Kumar, J.; Lacher, C.; Combs, G.F.; McComsey, G.A. Plasma Selenium Concentrations Are Sufficient and Associated with Protease Inhibitor Use in Treated HIV-Infected Adults. *J. Nutr.* **2015**, *145*, 2293–2299. [CrossRef] [PubMed]

216. Baeten, J.M.; Mostad, S.B.; Hughes, M.P.; Overbaugh, J.; Bankson, D.D.; Mandaliya, K.; Ndinya-Achola, J.O.; Bwayo, J.J.; Kreiss, J.K. Selenium deficiency is associated with shedding of HIV-1—Infected cells in the female genital tract. *J. Acquir. Immune Defic. Syndr.* **2001**, *26*, 360–364. [CrossRef] [PubMed]

217. Kupka, R.; Msamanga, G.I.; Xu, C.; Anderson, D.; Hunter, D.; Fawzi, W.W. Relationship between plasma selenium concentrations and lower genital tract levels of HIV-1 RNA and interleukin type 1beta. *Eur. J. Clin. Nutr.* **2007**, *61*, 542–547. [CrossRef] [PubMed]

218. Jones, C.Y.; Tang, A.M.; Forrester, J.E.; Huang, J.; Hendricks, K.M.; Knox, T.A.; Spiegelman, D.; Semba, R.D.; Woods, M.N. Micronutrient levels and HIV disease status in HIV-infected patients on highly active antiretroviral therapy in the Nutrition for Healthy Living cohort. *J. Acquir. Immune Defic. Syndr.* **2006**, *43*, 475–482. [CrossRef]

219. Kupka, R.; Mugusi, F.; Aboud, S.; Msamanga, G.I.; Finkelstein, J.L.; Spiegelman, D.; Fawzi, W.W. Randomized, double-blind, placebo-controlled trial of selenium supplements among HIV-infected pregnant women in Tanzania: Effects on maternal and child outcomes. *Am. J. Clin. Nutr.* **2008**, *87*, 1802–1808. [CrossRef] [PubMed]

220. Look, M.P.; Rockstroh, J.K.; Rao, G.S.; Barton, S.; Lemoch, H.; Kaiser, R.; Kupfer, B.; Sudhop, T.; Spengler, U.; Sauerbruch, T. Sodium selenite and N-acetylcysteine in antiretroviral-naive HIV-1-infected patients: A randomized, controlled pilot study. *Eur. J. Clin. Invest.* **1998**, *28*, 389–397. [CrossRef] [PubMed]

221. McClelland, R.S.; Baeten, J.M.; Overbaugh, J.; Richardson, B.A.; Mandaliya, K.; Emery, S.; Lavreys, L.; Ndinya-Achola, J.O.; Bankson, D.D.; Bwayo, J.J.; et al. Micronutrient supplementation increases genital tract shedding of HIV-1 in women: Results of a randomized trial. *J. Acquir. Immune Defic. Syndr.* **2004**, *37*, 1657–1663. [CrossRef]

222. Constans, J.; Delmas-Beauvieux, M.C.; Sergeant, C.; Peuchant, E.; Pellegrin, J.L.; Pellegrin, I.; Clerc, M.; Fleury, H.; Simonoff, M.; Leng, B.; et al. One-year antioxidant supplementation with beta-carotene or selenium for patients infected with human immunodeficiency virus: A pilot study. *Clin. Infect. Dis.* **1996**, *23*, 654–656. [CrossRef]

223. Durosinmi, M.A.; Armistead, H.; Akinola, N.O.; Onayemi, O.; Adediran, I.A.; Olasode, O.A.; Elujoba, A.A.; Irinoye, O.; Ogun, S.A.; Odusoga, O.L.; et al. Selenium and aspirin in people living with HIV and AIDS in Nigeria. *Niger. Postgrad. Med. J.* **2008**, *15*, 215–218.

224. Sudfeld, C.R.; Aboud, S.; Kupka, R.; Mugusi, F.M.; Fawzi, W.W. Effect of selenium supplementation on HIV-1 RNA detection in breast milk of Tanzanian women. *Nutrition (Burbank)* **2014**, *30*, 1081–1084. [CrossRef]

225. Kupka, R.; Mugusi, F.; Aboud, S.; Hertzmark, E.; Spiegelman, D.; Fawzi, W.W. Effect of selenium supplements on hemoglobin concentration and morbidity among HIV-1-infected Tanzanian women. *Clin. Infect. Dis. Off. Pub. Infect. Dis. Soc. Am.* **2009**, *48*, 1475–1478. [CrossRef]

226. Richard, M.J.; Guiraud, P.; Didier, C.; Seve, M.; Flores, S.C.; Favier, A. Human immunodeficiency virus type 1 Tat protein impairs selenoglutathione peroxidase expression and activity by a mechanism independent of cellular selenium uptake: Consequences on cellular resistance to UV-A radiation. *Arch. Biochem. Biophys.* **2001**, *386*, 213–220. [CrossRef] [PubMed]

227. Sappey, C.; Legrand-Poels, S.; Best-Belpomme, M.; Favier, A.; Rentier, B.; Piette, J. Stimulation of glutathione peroxidase activity decreases HIV type 1 activation after oxidative stress. *AIDS Res. Hum. Retrovir.* **1994**, *10*, 1451–1461. [CrossRef] [PubMed]

228. Makropoulos, V.; Bruning, T.; Schulze-Osthoff, K. Selenium-mediated inhibition of transcription factor NF-kappa B and HIV-1 LTR promoter activity. *Arch. Toxicol.* **1996**, *70*, 277–283. [CrossRef] [PubMed]

229. Hori, K.; Hatfield, D.; Maldarelli, F.; Lee, B.J.; Clouse, K.A. Selenium supplementation suppresses tumor necrosis factor alpha-induced human immunodeficiency virus type 1 replication in vitro. *AIDS Res. Hum. Retrovir.* **1997**, *13*, 1325–1332. [CrossRef] [PubMed]

230. Sandstrom, P.A.; Murray, J.; Folks, T.M.; Diamond, A.M. Antioxidant defenses influence HIV-1 replication and associated cytopathic effects. *Free Radic. Biol. Med.* **1998**, *24*, 1485–1491. [CrossRef]

231. Kalantari, P.; Narayan, V.; Natarajan, S.K.; Muralidhar, K.; Gandhi, U.H.; Vunta, H.; Henderson, A.J.; Prabhu, K.S. Thioredoxin reductase-1 negatively regulates HIV-1 transactivating protein Tat-dependent transcription in human macrophages. *J. Biol. Chem.* **2008**. [CrossRef] [PubMed]

232. Benelli, J.L.; de Medeiros, R.M.; Matte, M.C.C.; de Melo, M.G.; de Matos Almeida, S.E.; Fiegenbaum, M. Role of SEP15 Gene Polymorphisms in the Time of Progression to AIDS. *Gen. Test. Mol. Biomark.* **2016**, *20*, 383–387. [CrossRef]

233. Zhao, L.; Cox, A.G.; Ruzicka, J.A.; Bhat, A.A.; Zhang, W.; Taylor, E.W. Molecular modeling and in vitro activity of an HIV-1-encoded glutathione peroxidase. *Proc. Natl. Acad. Sci. USA* **2000**, *97*, 6356–6361. [CrossRef]

234. Cohen, I.; Boya, P.; Zhao, L.; Metivier, D.; Andreau, K.; Perfettini, J.L.; Weaver, J.G.; Badley, A.; Taylor, E.W.; Kroemer, G. Anti-apoptotic activity of the glutathione peroxidase homologue encoded by HIV-1. *Apoptosis* **2004**, *9*, 181–192. [CrossRef]

235. Xu, X.-M.; Carlson, B.A.; Grimm, T.A.; Kutza, J.; Berry, M.J.; Arreola, R.; Fields, K.H.; Shanmugam, I.; Jeang, K.-T.; Oroszlan, S.; et al. Rhesus monkey simian immunodeficiency virus infection as a model for assessing the role of selenium in AIDS. *J. Acquir. Immune Defic. Syndr. (1999)* **2002**, *31*, 453–463. [CrossRef]

236. Chen, C.; Zhou, J.; Xu, H.; Jiang, Y.; Zhu, G. Effect of selenium supplementation on mice infected with LP-BM5 MuLV, a murine AIDS model. *Biol. Trace Elem. Res.* **1997**, *59*, 187–193. [CrossRef] [PubMed]

237. Chen, X.; Anstey, A.V.; Bugert, J.J. Molluscum contagiosum virus infection. *Lancet Infect. Dis.* **2013**, *13*, 877–888. [CrossRef]

238. Schaffer, J.V.; Berger, E.M. Molluscum Contagiosum. *JAMA Dermatol.* **2016**, *152*, 1072. [CrossRef] [PubMed]

239. McFadden, G. Even viruses can learn to cope with stress. *Science* **1998**, *279*, 40–41. [CrossRef] [PubMed]

240. Mendez-Rios, J.D.; Yang, Z.; Erlandson, K.J.; Cohen, J.I.; Martens, C.A.; Bruno, D.P.; Porcella, S.F.; Moss, B. Molluscum Contagiosum Virus Transcriptome in Abortively Infected Cultured Cells and a Human Skin Lesion. *J. Virol.* **2016**, *90*, 4469–4480. [CrossRef] [PubMed]

241. Taylor, E.W.; Bhat, A.; Nadimpalli, R.G.; Zhang, W.; Kececioglu, J. HIV-1 encodes a sequence overlapping env gp41 with highly significant similarity to selenium-dependent glutathione peroxidases. *J. Acquir. Immune Def. Syndr. Hum. Retrovirol. Off. Pub. Int. Retrovirol. Assoc.* **1997**, *15*, 393–394. [CrossRef]

Review

Role of Selenoprotein F in Protein Folding and Secretion: Potential Involvement in Human Disease

Bingyu Ren [1,2], **Min Liu** [1,3], **Jiazuan Ni** [3,4] **and Jing Tian** [1,3,4,*]

[1] Shenzhen Key Laboratory of Marine Biotechnology and Ecology, Department of Marine Biology, Shenzhen University, Shenzhen 518060, China; renbingyu89@foxmail.com (B.R.); 2170257211@email.szu.edu.cn (M.L.)

[2] Changchun Institute of Applied Chemistry, Chinese Academy of Sciences, Changchun 130022, China

[3] Shenzhen Engineering Laboratory for Marine Algal Biotechnology, College of Life Sciences and Oceanography, Shenzhen University, Shenzhen 518060, China; jzni@szu.edu.cn

[4] Shenzhen Key Laboratory of Microbial Genetic Engineering, College of Life Sciences and Oceanography, Shenzhen University, Shenzhen 518060, China

* Correspondence: jtian@szu.edu.cn; Tel.: +86-755-8671-3951

Received: 29 September 2018; Accepted: 23 October 2018; Published: 2 November 2018

Abstract: Selenoproteins form a group of proteins of which its members contain at least one selenocysteine, and most of them serve oxidoreductase functions. Selenoprotein F (SELENOF), one of the 25 currently identified selenoproteins, is located in the endoplasmic reticulum (ER) organelle and is abundantly expressed in many tissues. It is regulated according to its selenium status, as well as by cell stress conditions. SELENOF may be functionally linked to protein folding and the secretion process in the ER. Several studies have reported positive associations between *SELENOF* genetic variations and several types of cancer. Also, altered expression levels of SELENOF have been found in cancer cases and neurodegenerative diseases. In this review, we summarize the current understanding of the structure, expression, and potential function of SELENOF and discuss its possible relation with various pathological processes.

Keywords: selenium; selenoprotein F; thiol–disulfide oxidoreductase; endoplasmic reticulum stress; protein folding quality control; single nucleotide polymorphisms

---

1. Background

As a trace element, selenium is nutritionally essential for mammals. A lot of evidence has revealed the role of selenium in preventing cancers and maintaining the proper function of the thyroid, the immune system, and reproduction [1]. Selenium deficiency is connected to cardiovascular, aging-associated, immune, and brain diseases [2–5]. Nevertheless, selenium has been found to have a two-sided effect on human health, depending on its concentration. Excessive selenium can be toxic, and a tolerable upper intake level of selenium has been established at 300 μg/person per day in the EU [6]. Also, a U-shaped association has been noted between selenium status and its health effects [7].

The functions of selenium are believed to be enabled by the 21st amino acid, selenocysteine (Sec, U, Se-Cys) [8]. Sec is encoded by the TGA codon, which is recognized as a stop codon in general cases. Specific complex machinery is required to co-translationally incorporate Sec into the growing polypeptide chain, including selenocysteine insertion (SECIS) elements in the selenoprotein mRNA secondary structures and trans-acting factors that interact with SECIS elements [9].

Proteins containing Sec are grouped into the so-called selenoproteins, which function as regulators, structural proteins, essential antioxidant enzymes, and others [10]. The protective effect of selenium is achieved through the regulation of the expression of selenoproteins. Twenty-five selenoproteins have been discovered in the human proteome [11]. Selenoprotein F (SELENOF), the new name according to the selenoprotein gene nomenclature, was initially named the 15-kDa selenoprotein (Sep15) [12]. It was first characterized in human T-cells by Gladyshev in 1998. A selenocysteine residue was found in the polypeptide sequences of SELENOF, together with a SECIS element that was located in the 3′-UTR of its mRNA, thus meeting the criteria of a selenoprotein [13]. Though the function of SELENOF is only partially established so far, its potential roles in protein folding and secretion have been indicated. Here, we summarize the available knowledge relating to the structure, distribution, and regulation of SELENOF, explore the role of SELENOF in protein redox quality control, and discuss the possible associations between *SELENOF* genetic polymorphisms/dysregulation and pathologies.

2. The Cellular Localization and Structure of SELENOF

The human *SELENOF* gene is located on chromosome 1p31 [14]. The gene product SELENOF is a protein that is localized in the endoplasmic reticulum (ER) with a molecular mass that is close to 15 kDa. It has been found in various species, from green algae to humans. Alignment results have revealed that its sequence is highly conserved and shares 31% similarity with another ER selenoprotein, Selenoprotein M [15].

According to the relative location of Sec in the polypeptide chain, selenoproteins can be divided into two groups. One group contains Sec close to the C-terminal, and the other group has Sec in the N-terminal part, which has a thioredoxin (Trx)-like structure in most cases [11,16]. SELENOF belongs to the latter group. Structural studies have revealed an ER signal peptide at its N-terminus, which directs the newly synthesized SELENOF to the ER where it will be cleaved into its mature form. Though no typical ER-resident peptide was found in the SELENOF sequence, it is maintained in ER through a tight association with another protein, UDP-glucose:glycoprotein glucosyltransferase (UGGT), through a cysteine-rich domain [17]. An atypical CxU motif, in which Cys is separated from Sec by only one amino acid, is located in the Trx-like domain at the C-terminus [18]. Since Sec is usually present as an active-site residue and the CxU motif of SELENOF is redox-active and surface accessible, catalytic activity can be expected from SELENOF. The equilibrium redox potential of the fruit fly SELENOF protein has been measured to be −225 mV, which is right between the redox potentials of Trx and protein disulfide isomerase (PDI), suggesting that SELENOF is capable of reducing and/or isomerizing the disulfide bonds of proteins [18].

3. The Expression and Regulation of SELENOF

The ER is the major cellular component for secreted protein manufacture and glycosylation initiation. As an ER protein, SELENOF can be detected in a wide variety of human tissues. It is notably expressed in tissues with secretory functions, such as thyroid, liver, and kidney; in reproductive organs, such as prostate and testis [14]; and in abundance in brain regions like the hippocampus and cerebellum [19]. In silicopredictions of the human *SELENOF* gene have found a typical CpG island (regions with a high frequency of CpG siteswhere a cytosine nucleotide is followed by a guanine nucleotide), two putative metal response elements (MREs), and four putativenuclear factor kappa-light-chain-enhancer of activated B cells (NFκB) binding sites located upstream of its transcription start site [20] which may mediate the tissue-specific transcriptional expression of SELENOF. Our research group has confirmed that NFκB mediates the transcriptional regulation of SELENOF expression in HEK293T cells [21]. Interestingly, MREs are features of the binding sequences of metal regulatory transcription factors, a class of transcription factors that regulate the transcriptional response to heavy metal exposure, oxidative stress, and hypoxia. The possibility that metal regulatory transcription factors regulate SELENOF expression needs to be further clarified.

Hierarchical principles of selenoprotein expression in response to selenium have been noted by researchers using different tissues or cell models [22,23]. Generally, selenoproteins have been classified into two groups: "house-keeping" selenoproteins that are resistant to selenium changes, and "stress-regulated" selenoproteins that are sensitive to selenium changes [24]. Ina recently published study, SELENOF was assigned to the selenium-sensitive "stress-regulated" group [25]. The expression of SELENOF has been reported to be regulated by selenium bothin vitro and in vivo. In primary mouse colon cells, both selenite and methylseleninic acid (MeSeA) increased SELENOF protein levels in a dose-dependent manner, while the very same concentrations of Se-methyl selenocysteine (SeMeSeCys) and selenomethionine (SeMet) had no effect. Notably, the mRNA levels of SELENOF remained unaffected by all four of these selenium compounds, indicating that the increase in SELENOF protein levels may result fromthe translation of still-present mRNA [26]. In chickens, dietary selenium deficiency led to a decrease inSELENOF mRNA [27]. In the liver of selenium-supplemented growing lambs, SELENOF mRNA was found to be upregulated [28]. Furthermore, selenium can prevent the downregulation of SELENOF mRNA which is caused by toxicants or oxidative stress [29–31].

Beyond that, the expression of SELENOF is also connected with the state of the ER. Different changing tendencies of SELENOF protein levels have been observed in adaptive and acute ER stress conditions [19]. The pharmacological ER stress inducers Tunicamycin and Brefeldin A increased SELENOF during a 24 h treatment, while Thapsigargin and Dithiothreitol (DTT) stimulated the rapid degradation of SELENOF [19]. Notably, both Tunicamycin and Brefeldin A trigger ER stress by interfering with protein synthesis, which indicates the involvement of SELENOF during this process.

4. Role of SELENOF in Redox Protein Quality Control

It has been recognized by researchers that the antioxidative and stress-relieving effects of selenium are mainly achieved by its incorporation into the selenoproteins with oxidoreductase functions [32]. During adaptive ER stress, SELENOF has been found to be upregulated with a series of protective cellular actions, such as the Unfolded Protein Response [19]. Thus, the protective function of SELENOF can be assumed to occur during the selenium-mediated antioxidative process and under ER stress conditions.

Seven selenoproteins have been identified as residents of the ER, including Iodothyronine deiodinase 2 (DIO2) and selenoproteins F, K, M, N, S, and T. Though the functions of these ER-resident selenoproteins have not been fully characterized, evidence has indicated their involvement in several ER processes, such as maintaining redox or calcium homeostasis, quality control, and the endoplasmic-reticulum-associated protein degradation (ERAD) machinery [33]. As SELENOF's binding partner, UGGT participates in ER protein folding quality control by reglucosylating misfolded glycoproteins [34]. SELENOF has been found to form a heterodimeric complex with both UGGT isoforms and markedly enhance their glucosyltransferase activity [35]. The UGGT1 knockout is embryo lethal, revealing that it is indispensable for embryogenesis and maintaining tissue function [36]. Knockout studies of UGGT1 in cultured cells have demonstrated its effect on protein solubility and secretion rates [37,38]. Since structural studies have indicated that SELENOF may exhibit thiol–disulfide oxidoreductase activity, it may also play a role in the glycoprotein folding process of misfolded proteins that are recognized by UGGT as its potential substrates [15], suggesting that SELENOF deficiency possibly results in similar outcomes to the UGGT knockout.

We summarize in Table 1 the main contents of the SELENOF knockdown or knockout studies that have been reported so far. SELENOF knockout mice are viable and fertile, so observation of abnormal protein remodeling or protein secretion have been reported in SELENOF deficiency cases. In the latest work on knockout mice, a nonfunctional increase in the secretion of the disulfide-rich glycoprotein Immunoglobulin M (IgM) and a delay of ER–Golgi protein transportation were detected. Based on this result, together with previous reports, a gatekeeper function of SELENOF in the redox quality control process of glycoprotein secretion has been proposed. Recently, we identified Retinol dehydrogenase 11 (RDH11) as an interacting protein for SELENOF. SELENOF overexpression caused a decrease in exogenous RDH11 retinal reductase activity, suggesting that SELENOF may also affect protein enzyme activities [39].

Table 1. Summarization of the SELENOF (Selenoprotein F) knockdown and knockout studies.

No.	Models	Methods	Phenotypes	Pathways or Biological Processes that Were Possibly Involved	Published Year
1	Mouse malignant mesothelioma (MM) cells	siRNA	The expression of SELENOF was downregulated in most MM cases. Differential effects of selenium on MM cell growth were associated with genotype and expression of SELENOF.	The selenium-induced MM cell apoptosis was increased in cells that were transfected with wild-type SELENOF, however not with the 1125A variant. SELENOF siRNA inhibition made the sensitive MM cells more resistant to selenium [40].	2004
2	Mouse CT26 colon cancer cells and Lewis lung carcinoma (LLC1) lung cancer cells	Stably transfected shRNA	Tumorigenicity and metastasis inhibition together with G2/M cell cycle arrest in colon cancer cells; no effect on lung cancer cells.	Genes significantly affected by SELENOF downregulation belonged to cancer, cellular growth, and proliferation biological processes [41].	2010
3	Mouse	Knockout	Mice were viable and fertile, with normal brain morphology and no activation of endoplasmic reticulum(ER) stress. The oxidative stress was elevated in the livers, and prominent nuclear cataracts were developing at an early age.	SELENOF mRNA level was progressively elevated in the lens during mouse development. An improper folding status of lens proteins was possibly caused by SELENOF deficiency [42].	2011
4	Mouse	Knockout	Protected mice against chemically induced colon cancer by inhibiting aberrant crypt formation.	SELENOF knockout resulted in upregulation of Guanylate binding protein-1 mRNA and protein expression and a higher level of interferon-γ in plasma [43].	2012
5	Chang liver cells	Doxycycline-inducible shRNA	Actin and Tubulin cytoskeleton protein remodeling and non-apoptotic membrane blebbing.	SELENOF knockdown induced Ras homolog gene family, member A (RhoA) activation and phosphorylation of myosin phosphatase target subunit 1, and the remodeling of F-actin and α-tubulin was different from typical apoptotic blebbing cells [44].	2015
6	Chang liver cells	Doxycycline-inducible shRNA	Cell proliferation and motility inhibition together with G1 cell cycle arrest.	Activation of ER stress, upregulation of p21 and p27, and relocation of focal adhesions in SELENOF-deficient cells [45].	2015
7	Human Lens Epithelial cells	siRNA	Aggravation of the tunicamycin-induced cell apoptosis.	SELENOF knockdown further exacerbated Caspase activation, mitochondrial membrane potential decrease, cytochrome C release, and reactive oxygen species (ROS) generation, with no effect on ER stress [46].	2015
8	Mouse CT26 colon cancer cells	Stably transfected shRNA	Growth and metastasis inhibition in either SELENOF or thioredoxin reductase 1 downregulated colon cancer cells.	Inflammation-related genes regulated by Stat-1, especially interferon-γ-regulated guanylate-binding proteins, were highly elevated in SELENOF-deficient cells, however not in thioredoxin reductase 1-deficient cells. Wnt/β-catenin signaling pathway was upregulated in cells lacking both thioredoxin reductase 1 and SELENOF [47].	2015
9	Mouse	Knockout	Mild splenomegaly and elevated Immunoglobulin Levels without altering immune functions.	Increased secretion of Immunoglobulin M (IgM), delay of ER-to-Golgi glycoprotein transportation [48].	2018

5. Associations of SELENOF and Disease Pathologies

Connections have been noted between selenium status and many disease progressions, such as various cancers, immune system diseases, and neurodegenerative diseases [32,49,50]. Though supplementation with selenium has obtained some beneficial effects during these cases, current findings are not conclusive enough and are sometimes inconsistent. For example, both the Selenium and Vitamin E Cancer Prevention Trial (SELECT) and a Phase III trial of selenium have failed to confirm the effective role of selenium in prostate cancer prevention from the Nutritional Prevention of Cancer Trial (NPC) results [51]. The disparate results of these trials may be explained by the U-shaped relationship between selenium status and protection from cancer [7].

Besides selenium status, attention has also been paid to the possible roles of selenoproteins in these pathologies. Evidence for the effect of SELENOF deficiency on several different cellular events has been reported, together with the abundant expression of SELENOF in various organs, which we reviewed above, suggesting that SELENOF could be associated with multiple pathological processes in different tissues.

5.1. SELENOF Gene Polymorphisms and Pathologies

SELENOF genetic variations have been found to be associated with cancer etiology. Two single-nucleotide polymorphisms (SNPs), rs5845 and rs5859, have been identified in its 3′-untranslated region within the selenocysteine insertion sequence-like structures [14]. A significant association between the rs5845 T allele variant and elevated breast cancer risk in African-American women has been observed [14]. However, the same genotype shows no association with breast cancer risk or clinicopathological parameters in Caucasian women [52]. A significant difference in genotype distribution of the other site of polymorphism, rs5859, has been found between breast cancer patients and controls in the population of Iran [53]. In smoking individuals, the rs5859 AA genotype may still benefit from selenium when its plasma concentration is higher than 80 ng/mL, whereas in those with the GG or GA genotype, a relatively high selenium status could increase the risk of lung cancer [54]. Both the rs5845 and rs5859 sites with minor A and T alleles are associated with increased risk of male rectal cancer in the Korean population, which suggests that the effect of the two SNPs on cancer may be gender-dependent [55]. Another four common SNPs within the *SELENOF* gene—rs479341, rs527281, rs561104, and rs1407131—have also been identified. Though the polymorphisms rs479341, rs1407131, and rs561104 are not significantly associated with prostate cancer risk, these SNPs are significantly associated with prostate cancer mortality [56].

In addition, connections have been found between *SELENOF* gene polymorphisms and other diseases or functions. In Kashin–Beck disease, the minor Aallele frequency of rs5859 is statistically significantly higher [57]. This AA genotype at the rs5859 site is also associated with a shorter time of progression to AIDS compared with GG homozygotes [58]. In volunteers that aremore than 50 years old, rs5845 C allele variants received higher verbal learning memory scores than T allele variants [59]. The polymorphic sequences of rs5845 and rs5859 can alter the Sec incorporation efficiency of SELENOF SECIS elements [14], indicating that *SELENOF* genetic polymorphism may lead to different SELENOF protein expression levels in response to selenium, which might contribute to these disease pathologies.

5.2. SELENOF Dysregulation in Pathologies

Dysregulation of SELENOF at mRNA levels has also been noted in several disease cases or pathological models. SELENOF was found to be upregulated in two hepatocellular carcinoma cell lines, HepG2 and Huh7, compared with normal human hepatocytes [60]. On the contrary, it was downregulated in almost 60% of the malignant mesothelioma cell lines and tumor specimens compared with normal mesothelial cells [40]. Also, the expression of SELENOF mRNA was downregulated in the hippocampus and substantia nigra brain regions of a Parkinson's mouse model [61] and in the leukocytes of bladder cancer patients [62].

6. Conclusions and Perspectives

Sec is encoded by the genetic code UGA, which is recognized as a stop codon in general cases. It is difficult to overexpress the wild-type selenoproteins in cell or animal models, which limits the approaches for SELENOF functional studies. Still, the understanding of SELENOF's unique structure, subcellular localizations, binding partner, tissue distribution, and regulation will give us some clues about its potential function. A series of deficiency studies have revealed its effect on protein secretion, supporting the speculation that SELENOF plays a role in the redox protein folding process in the ER.

As shown in the concluding Figure 1, as an ER selenoprotein, SELENOF can be regulated by both selenium status and ER stress. SELENOF forms a complex with UGGT and is involved in glycoprotein folding quality control. The newly synthesized and folded proteins are packed into vesicles, followed by Golgi transportation and secretion.

Accordingly, an increasing number of studies have linked *SELENOF* gene polymorphisms and SELENOF dysregulation to various diseases, including several types of cancer, AIDS, and neurodegeneration, which reveals the importance of SELENOF's physiological functions. Currently, the mechanisms underlying SELENOF's associations with numerous pathological states are not fully understood, and additional work is required to confirm SELENOF's role in protein quality control. Recently, we applied a biotin labeling method to the screening and identification of SELENOF's potential substrates (data unpublished) which may help us to better understand SELENOF's involvement in various cellular processes. Promising results have been reported in cases applying selenium compounds to cancer and neurodegeneration treatment. Since SELENOF is selenium-sensitive, it may be a noteworthy potential target during these pathologies.

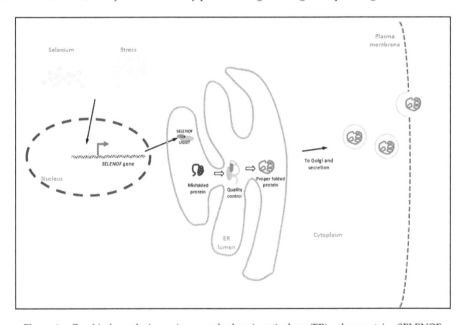

Figure 1. Graphical conclusion. As an endoplasmic reticulum (ER) selenoprotein, SELENOF can be regulated by both selenium status and ER stress. It forms a 1:1 tight complex with UDP-glucose:glycoprotein glucosyltransferase (UGGT) in the ER lumen, thus enhancing the enzymatic activity of UGGT. SELENOF may be involved in glycoprotein folding quality control by rearranging or reducing the disulfide bonds of UGGT-recognized misfolded proteins. The properly folded proteins are then packed into vesicles, followed by Golgi transportation and secretion.

Author Contributions: B.R. conceived and led the writing of the manuscript. M.L. helped to collect and sort the related literature. J.N. and J.T. critically revised the manuscript. All authors read and approved the final version of the manuscript.

Funding: This study was financially supported by grants from the National Natural Science Foundation of China (81372984) and the Science and Technology Innovation Committee of Shenzhen Municipality (JCYJ20170817094634747) awarded to J.T., as well as grants from the China Postdoctoral Science Foundation (2016M602531), the Natural Science Foundation of Guangdong Province (2017A030310140) awarded to B.R.

Conflicts of Interest: The authors declare that they have no competing interests.

References

1. Rayman, M.P. The importance of selenium to human health. *Lancet* **2000**, *356*, 233–241. [CrossRef]
2. Benstoem, C.; Goetzenich, A.; Kraemer, S.; Borosch, S.; Manzanares, W.; Hardy, G.; Stoppe, C. Selenium and its supplementation in cardiovascular disease—What do we know? *Nutrients* **2015**, *7*, 3094–3118. [CrossRef] [PubMed]
3. McCann, J.C.; Ames, B.N. Adaptive dysfunction of selenoproteins from the perspective of the triage theory: Why modest selenium deficiency may increase risk of diseases of aging. *FASEB J.* **2011**, *25*, 1793–1814. [CrossRef] [PubMed]
4. Khoso, P.A.; Yang, Z.; Liu, C.; Li, S. Selenium deficiency downregulates selenoproteins and suppresses immune function in chicken thymus. *Biol. Trace Elem. Res.* **2015**, *167*, 48–55. [CrossRef] [PubMed]
5. Pillai, R.; Uyehara-Lock, J.H.; Bellinger, F.P. Selenium and selenoprotein function in brain disorders. *IUBMB Life* **2014**, *66*, 229–239. [CrossRef] [PubMed]
6. European Commission Health & Consumer Protection Directorate-General Scientific Committee on Food. *Opinion of the Scientific Committee on Food on the Tolerable Upper Intake Level of Selenium*; European Commission Health & Consumer Protection Directorate-General Scientific Committee on Food: Brussels, Belgium, 2000.
7. Rayman, M.P. Selenium and human health. *Lancet* **2012**, *379*, 1256–1268. [CrossRef]
8. Low, S.C.; Berry, M.J. Knowing when not to stop: Selenocysteine incorporation in eukaryotes. *Trends Biochem. Sci.* **1996**, *21*, 203–208. [CrossRef]
9. Squires, J.E.; Berry, M.J. Eukaryotic selenoprotein synthesis: Mechanistic insight incorporating new factors and new functions for old factors. *IUBMB Life* **2008**, *60*, 232–235. [CrossRef] [PubMed]
10. Weeks, B.S.; Hanna, M.S.; Cooperstein, D. Dietary selenium and selenoprotein function. *Med. Sci. Monit.* **2012**, *18*, RA127–RA132. [CrossRef] [PubMed]
11. Kryukov, G.V.; Castellano, S.; Novoselov, S.V.; Lobanov, A.V.; Zehtab, O.; Guigo, R.; Gladyshev, V.N. Characterization of mammalian selenoproteomes. *Science* **2003**, *300*, 1439–1443. [CrossRef] [PubMed]
12. Gladyshev, V.N.; Arner, E.S.; Berry, M.J.; Brigelius-Flohe, R.; Bruford, E.A.; Burk, R.F.; Carlson, B.A.; Castellano, S.; Chavatte, L.; Conrad, M.; et al. Selenoprotein gene nomenclature. *J. Biol. Chem.* **2016**, *291*, 24036–24040. [CrossRef] [PubMed]
13. Gladyshev, V.N.; Jeang, K.T.; Wootton, J.C.; Hatfield, D.L. A new human selenium-containing protein. Purification, characterization, and cDNA sequence. *J. Biol. Chem.* **1998**, *273*, 8910–8915. [CrossRef] [PubMed]
14. Kumaraswamy, E.; Malykh, A.; Korotkov, K.V.; Kozyavkin, S.; Hu, Y.; Kwon, S.Y.; Moustafa, M.E.; Carlson, B.A.; Berry, M.J.; Lee, B.J.; et al. Structure-expression relationships of the 15-kDa selenoprotein gene. Possible role of the protein in cancer etiology. *J. Biol. Chem.* **2000**, *275*, 35540–35547. [CrossRef] [PubMed]
15. Labunskyy, V.M.; Hatfield, D.L.; Gladyshev, V.N. The Sep15 protein family: Roles in disulfide bond formation and quality control in the endoplasmic reticulum. *IUBMB Life* **2007**, *59*, 1–5. [CrossRef] [PubMed]
16. Lu, J.; Holmgren, A. Selenoproteins. *J. Biol. Chem.* **2009**, *284*, 723–727. [CrossRef] [PubMed]
17. Korotkov, K.V.; Kumaraswamy, E.; Zhou, Y.; Hatfield, D.L.; Gladyshev, V.N. Association between the 15-kDa selenoprotein and UDP-glucose: Glycoprotein glucosyltransferase in the endoplasmic reticulum of mammalian cells. *J. Biol. Chem.* **2001**, *276*, 15330–15336. [CrossRef] [PubMed]
18. Ferguson, A.D.; Labunskyy, V.M.; Fomenko, D.E.; Arac, D.; Chelliah, Y.; Amezcua, C.A.; Rizo, J.; Gladyshev, V.N.; Deisenhofer, J. NMR structures of the selenoproteins Sep15 and SelM reveal redox activity of a new thioredoxin-like family. *J. Biol. Chem.* **2006**, *281*, 3536–3543. [CrossRef] [PubMed]
19. Labunskyy, V.M.; Yoo, M.H.; Hatfield, D.L.; Gladyshev, V.N. Sep15, a thioredoxin-like selenoprotein, is involved in the unfolded protein response and differentially regulated by adaptive and acute ER stresses. *Biochemistry* **2009**, *48*, 8458–8465. [CrossRef] [PubMed]
20. Stoytcheva, Z.R.; Berry, M.J. Transcriptional regulation of mammalian selenoprotein expression. *Biochim. Biophys. Acta* **2009**, *1790*, 1429–1440. [CrossRef] [PubMed]
21. Zou, C.; Zheng, J.; Li, J.; Tian, J. Human gene of 15-kDa selenoprotein is regulated by NF κB. In *Selenium in the Environment and Human Health*; Gary, S., Banuelos, Z.-Q.L., Yin, X., Eds.; CRC Press: Boca Raton, FL, USA, 2013; pp. 172–173.

22. Behne, D.; Hilmert, H.; Scheid, S.; Gessner, H.; Elger, W. Evidence for specific selenium target tissues and new biologically important selenoproteins. *Biochim. Biophys. Acta* **1988**, *966*, 12–21. [CrossRef]

23. Sunde, R.A. Selenoproteins: Hierarchy, requirements, and biomarkers. In *Selenium: Its Molecular Biology and Role in Human Health*; Hatfield, D.L., Berry, M.J., Gladyshev, V.N., Eds.; Springer: New York, NY, USA, 2012; pp. 137–152.

24. Berry, M.J. Insights into the hierarchy of selenium incorporation. *Nat. Genet.* **2005**, *37*, 1162. [CrossRef] [PubMed]

25. Touat-Hamici, Z.; Bulteau, A.L.; Bianga, J.; Jean-Jacques, H.; Szpunar, J.; Lobinski, R.; Chavatte, L. Selenium-regulated hierarchy of human selenoproteome in cancerous and immortalized cells lines. *Biochim. Biophys. Acta* **2018**, *1862*, 2493–2505. [CrossRef] [PubMed]

26. Kipp, A.P.; Frombach, J.; Deubel, S.; Brigelius-Flohe, R. Selenoprotein was biomarker for the efficacy of selenium compounds to act as source for selenoprotein biosynthesis. In *Hydrogen Peroxide and Cell Signaling*; Pt, B., Cadenas, E., Packer, L., Eds.; Elsevier Academic Press Inc.: San Diego, CA, USA, 2013; Volume 527, pp. 87–112.

27. Zhang, J.L.; Xu, B.; Huang, X.D.; Gao, Y.H.; Chen, Y.; Shan, A.S. Selenium deficiency affects the mRNA expression of inflammatory factors and selenoprotein genes in the kidneys of broiler chicks. *Biol. Trace Elem. Res.* **2016**, *171*, 201–207. [CrossRef] [PubMed]

28. Juszczuk-Kubiak, E.; Bujko, K.; Cymer, M.; Wicinska, K.; Gabryszuk, M.; Pierzchala, M. Effect of inorganic dietary selenium supplementation on selenoprotein and lipid metabolism gene expression patterns in liver and loin muscle of growing lambs. *Biol. Trace Elem. Res.* **2016**, *172*, 336–345. [CrossRef] [PubMed]

29. Penglase, S.; Hamre, K.; Ellingsen, S. Selenium prevents downregulation of antioxidant selenoprotein genes by methylmercury. *Free Radic. Biol. Med.* **2014**, *75*, 95–104. [CrossRef] [PubMed]

30. Gao, H.; Liu, C.P.; Song, S.Q.; Fu, J. Effects of dietary selenium against lead toxicity on mRNA levels of 25 selenoprotein genes in the cartilage tissue of broiler chicken. *Biol. Trace Elem. Res.* **2016**, *172*, 234–241. [CrossRef] [PubMed]

31. Dai, J.; Zhou, J.; Liu, H.; Huang, K. Selenite and ebselen supplementation attenuates D-galactose-induced oxidative stress and increases expression of SELR and SEP15 in rat lens. *J. Biol. Inorg. Chem.* **2016**, *21*, 1037–1046. [CrossRef] [PubMed]

32. Misra, S.; Boylan, M.; Selvam, A.; Spallholz, J.E.; Björnstedt, M. Redox-active selenium compounds—From toxicity and cell death to cancer treatment. *Nutrients* **2015**, *7*, 3536–3556. [CrossRef] [PubMed]

33. Shchedrina, V.A.; Zhang, Y.; Labunskyy, V.M.; Hatfield, D.L.; Gladyshev, V.N. Structure-function relations, physiological roles, and evolution of mammalian ER-resident selenoproteins. *Antioxid. Redox Signal.* **2010**, *12*, 839–849. [CrossRef] [PubMed]

34. Ito, Y.; Takeda, Y.; Seko, A.; Izumi, M.; Kajihara, Y. Functional analysis of endoplasmic reticulum glucosyltransferase (UGGT): Synthetic chemistry's initiative in glycobiology. *Semin. Cell Dev. Biol.* **2015**, *41*, 90–98. [CrossRef] [PubMed]

35. Takeda, Y.; Seko, A.; Hachisu, M.; Daikoku, S.; Izumi, M.; Koizumi, A.; Fujikawa, K.; Kajihara, Y.; Ito, Y. Both isoforms of human UDP-glucose: Glycoprotein glucosyltransferase are enzymatically active. *Glycobiology* **2014**, *24*, 344–350. [CrossRef] [PubMed]

36. Ruddock, L.W.; Molinari, M. N-glycan processing in ER quality control. *J. Cell Sci.* **2006**, *119*, 4373–4380. [CrossRef] [PubMed]

37. Solda, T.; Galli, C.; Kaufman, R.J.; Molinari, M. Substrate-specific requirements for UGT1-dependent release from calnexin. *Mol. Cell* **2007**, *27*, 238–249. [CrossRef] [PubMed]

38. Ferris, S.P.; Jaber, N.S.; Molinari, M.; Arvan, P.; Kaufman, R.J.; Gilmore, R. UDP-glucose: Glycoprotein glucosyltransferase (UGGT1) promotes substrate solubility in the endoplasmic reticulum. *Mol. Biol. Cell* **2013**, *24*, 2597–2608. [CrossRef] [PubMed]

39. Tian, J.; Liu, J.; Li, J.; Zheng, J.; Chen, L.; Wang, Y.; Liu, Q.; Ni, J. The interaction of selenoprotein F (SELENOF) with retinol dehydrogenase 11 (RDH11) implied a role of SELENOF in vitamin A metabolism. *Nutr. Metab.* **2018**, *15*, 7. [CrossRef] [PubMed]

40. Apostolou, S.; Klein, J.O.; Mitsuuchi, Y.; Shetler, J.N.; Poulikakos, P.I.; Jhanwar, S.C.; Kruger, W.D.; Testa, J.R. Growth inhibition and induction of apoptosis in mesothelioma cells by selenium and dependence on selenoprotein Sep15 genotype. *Oncogene* **2004**, *23*, 5032–5040. [CrossRef] [PubMed]

41. Irons, R.; Tsuji, P.A.; Carlson, B.A.; Ouyang, P.; Yoo, M.H.; Xu, X.M.; Hatfield, D.L.; Gladyshev, V.N.; Davis, C.D. Deficiency in the 15-kDa selenoprotein inhibits tumorigenicity and metastasis of colon cancer cells. *Cancer Prev. Res.* **2010**, *3*, 630–639. [CrossRef] [PubMed]

42. Kasaikina, M.V.; Fomenko, D.E.; Labunskyy, V.M.; Lachke, S.A.; Qiu, W.; Moncaster, J.A.; Zhang, J.; Wojnarowicz, M.W., Jr.; Natarajan, S.K.; Malinouski, M.; et al. Roles of the 15-kDa selenoprotein (Sep15) in redox homeostasis and cataract development revealed by the analysis of Sep 15 knockout mice. *J. Biol. Chem.* **2011**, *286*, 33203–33212. [CrossRef] [PubMed]

43. Tsuji, P.A.; Carlson, B.A.; Naranjo-Suarez, S.; Yoo, M.H.; Xu, X.M.; Fomenko, D.E.; Gladyshev, V.N.; Hatfield, D.L.; Davis, C.D. Knockout of the 15 kDa selenoprotein protects against chemically-induced aberrant crypt formation in mice. *PLoS ONE* **2012**, *7*, e50574. [CrossRef] [PubMed]

44. Bang, J.; Jang, M.; Huh, J.H.; Na, J.W.; Shim, M.; Carlson, B.A.; Tobe, R.; Tsuji, P.A.; Gladyshev, V.N.; Hatfield, D.L.; et al. Deficiency of the 15-kDa selenoprotein led to cytoskeleton remodeling and non-apoptotic membrane blebbing through a RhoA/ROCK pathway. *Biochem. Biophys. Res. Commun.* **2015**, *456*, 884–890. [CrossRef] [PubMed]

45. Bang, J.; Huh, J.H.; Na, J.W.; Lu, Q.; Carlson, B.A.; Tobe, R.; Tsuji, P.A.; Gladyshev, V.N.; Hatfield, D.L.; Lee, B.J. Cell proliferation and motility are inhibited by G1 phase arrest in 15-kDa selenoprotein-deficient Chang liver cells. *Mol. Cells* **2015**, *38*, 457–465. [CrossRef] [PubMed]

46. Yin, N.; Zheng, X.; Zhou, J.; Liu, H.; Huang, K. Knockdown of 15-kDa selenoprotein (Sep15) increases hLE cells' susceptibility to tunicamycin-induced apoptosis. *J. Biol. Inorg. Chem.* **2015**, *20*, 1307–1317. [CrossRef] [PubMed]

47. Tsuji, P.A.; Carlson, B.A.; Yoo, M.H.; Naranjo-Suarez, S.; Xu, X.M.; He, Y.; Asaki, E.; Seifried, H.E.; Reinhold, W.C.; Davis, C.D.; et al. The 15 kDa selenoprotein and thioredoxin reductase 1 promote colon cancer by different pathways. *PLoS ONE* **2015**, *10*, e0124487. [CrossRef] [PubMed]

48. Yim, S.H.; Everley, R.A.; Schildberg, F.A.; Lee, S.G.; Orsi, A.; Barbati, Z.R.; Karatepe, K.; Fomenko, D.E.; Tsuji, P.A.; Luo, H.R.; et al. Role of selenof as a gatekeeper of secreted disulfide-rich glycoproteins. *Cell Rep.* **2018**, *23*, 1387–1398. [CrossRef] [PubMed]

49. Huang, Z.; Rose, A.H.; Hoffmann, P.R. The role of selenium in inflammation and immunity: From molecular mechanisms to therapeutic opportunities. *Antioxid. Redox Signal.* **2012**, *16*, 705–743. [CrossRef] [PubMed]

50. Chen, J.; Berry, M.J. Selenium and selenoproteins in the brain and brain diseases. *J. Neurochem.* **2003**, *86*, 1–12. [CrossRef] [PubMed]

51. Yang, L.; Pascal, M.; Wu, X.H. Review of selenium and prostate cancer prevention. *Asian Pac. J. Cancer Prev.* **2013**, *14*, 2181–2184. [CrossRef] [PubMed]

52. Watrowski, R.; Castillo-Tong, D.C.; Fabjani, G.; Schuster, E.; Fischer, M.; Zeillinger, R. The 811 c/t polymorphism in the 3′ untranslated region of the selenoprotein 15-kDa (Sep15) gene and breast cancer in caucasian women. *Tumour Biol.* **2016**, *37*, 1009–1015. [CrossRef] [PubMed]

53. Mohammaddoust, S.; Salehi, Z.; Saeidi Saedi, H. *SEPP1* and *SEP15* gene polymorphisms and susceptibility to breast cancer. *Br. J. Biomed. Sci.* **2018**, *75*, 36–39. [CrossRef] [PubMed]

54. Jablonska, E.; Gromadzinska, J.; Sobala, W.; Reszka, E.; Wasowicz, W. Lung cancer risk associated with selenium status is modified in smoking individuals by Sep15 polymorphism. *Eur. J. Nutr.* **2008**, *47*, 47–54. [CrossRef] [PubMed]

55. Sutherland, A.; Kim, D.H.; Relton, C.; Ahn, Y.O.; Hesketh, J. Polymorphisms in the selenoprotein S and 15-kDa selenoprotein genes are associated with altered susceptibility to colorectal cancer. *Genes Nutr.* **2010**, *5*, 215–223. [CrossRef] [PubMed]

56. Penney, K.L.; Schumacher, F.R.; Li, H.; Kraft, P.; Morris, J.S.; Kurth, T.; Mucci, L.A.; Hunter, D.J.; Kantoff, P.W.; Stampfer, M.J.; et al. A large prospective study of Sep15 genetic variation, interaction with plasma selenium levels, and prostate cancer risk and survival. *Cancer Prev. Res.* **2010**, *3*, 604–610. [CrossRef] [PubMed]

57. Wu, R.; Zhang, R.; Xiong, Y.; Sun, W.; Li, Y.; Yang, X.; Liu, J.; Jiang, Y.; Guo, H.; Mo, X.; et al. The study on polymorphisms of sep15 and TrxR2 and the expression of AP-1 signaling pathway in Kashin-Beck disease. *Bone* **2018**. [CrossRef] [PubMed]

58. Benelli, J.L.; de Medeiros, R.M.; Matte, M.C.; de Melo, M.G.; de Matos Almeida, S.E.; Fiegenbaum, M. Role of Sep15 gene polymorphisms in the time of progression to aids. *Genet. Test. Mol. Biomark.* **2016**, *20*, 383–387. [CrossRef] [PubMed]

59. Da Rocha, T.J.; Blehm, C.J.; Bamberg, D.P.; Fonseca, T.L.; Tisser, L.A.; de Oliveira Junior, A.A.; de Andrade, F.M.; Fiegenbaum, M. The effects of interactions between selenium and zinc serum concentration and Sep15 and SLC30A3 gene polymorphisms on memory scores in a population of mature and elderly adults. *Genes Nutr.* **2014**, *9*, 377. [CrossRef] [PubMed]

60. Guariniello, S.; Di Bernardo, G.; Colonna, G.; Cammarota, M.; Castello, G.; Costantini, S. Evaluation of the selenotranscriptome expression in two hepatocellular carcinoma cell lines. *Anal. Cell. Pathol.* **2015**, *2015*, 419561. [CrossRef] [PubMed]

61. Zhang, X.; Ye, Y.L.; Zhu, H.; Sun, S.N.; Zheng, J.; Fan, H.H.; Wu, H.M.; Chen, S.F.; Cheng, W.H.; Zhu, J.H. Selenotranscriptomic analyses identify signature selenoproteins in brain regions in a mouse model of parkinson's disease. *PLoS ONE* **2016**, *11*, e0163372. [CrossRef] [PubMed]

62. Reszka, E.; Gromadzinska, J.; Jablonska, E.; Wasowicz, W.; Jablonowski, Z.; Sosnowski, M. Level of selenoprotein transcripts in peripheral leukocytes of patients with bladder cancer and healthy individuals. *Clin. Chem. Lab. Med.* **2009**, *47*, 1125–1132. [CrossRef] [PubMed]

MDPI

St. Alban-Anlage 66

4052 Basel

Switzerland

Tel. +41 61 683 77 34

Fax +41 61 302 89 18

www.mdpi.com

Nutrients Editorial Office

E-mail: nutrients@mdpi.com

www.mdpi.com/journal/nutrients

Lightning Source UK Ltd.
Milton Keynes UK
UKHW050611040123
414766UK00004B/126